The Essence of Eternity:
Quaternion & Octonion SuperStandard Theories

Stephen Blaha Ph. D.
Blaha Research

Pingree-Hill Publishing
MMXX

To Margaret

Some Other Books by Stephen Blaha

All the Megaverse! Starships Exploring the Endless Universes of the Cosmos using the Baryonic Force (Blaha Research, Auburn, NH, 2014)

SuperCivilizations: Civilizations as Superorganisms (McMann-Fisher Publishing, Auburn, NH, 2010)

All the Universe! Faster Than Light Tachyon Quark Starships & Particle Accelerators with the LHC as a Prototype Starship Drive Scientific Edition (Pingree-Hill Publishing, Auburn, NH, 2011).

Unification of God Theory and Unified SuperStandard Model THIRD EDITION (Pingree Hill Publishing, Auburn, NH, 2018).

The Exact QED Calculation of the Fine Structure Constant Implies ALL 4D Universes have the Same Physics/Life Prospects (Pingree Hill Publishing, Auburn, NH, 2019).

Unified SuperStandard Theory and the SuperUniverse Model: The Foundation of Science (Pingree Hill Publishing, Auburn, NH, 2018).

Quaternion Unified SuperStandard Theory (The QUeST) and Megaverse Octonion SuperStandard Theory (MOST) (Pingree Hill Publishing, Auburn, NH, 2020).

United Universes Quaternion Universe - Octonion Megaverse (Pingree Hill Publishing, Auburn, NH, 2020).

Available on Amazon.com, bn.com Amazon.co.uk and other international web sites as well as at better bookstores (through Ingram Distributors).

CONTENTS

FIGURES and TABLES

ines on the left side (only shown for one layer) display the Generation mixing within each layer's species. The Generation mixing applies within each layer using a separate Generation group for each layer. The lines on the right side show Layer group mixing with the mixing amongst all four layers for each of the four generations individually. There are four Layer groups. The Dark groups mixing between normal and Dark fermions are shown in the center as horizontal lines. There are 256 fundamental fermions counting quarks as triplets. 32

INTRODUCTION

In previous books this author has derived the Unified SuperStandard Theory (UST) in our 3 + 1 dimension space-time from Complex General Relativity and Quantum Field Theory suitably extended. This book expands on Blaha (2020c) to give a more detailed derivation of UST from a 32 dimension complex quaternion space. Remarkably the internal symmetries of UST follow directly from the complex quaternion space discussion.

Similarly it also defines a complex octonion space of 32 dimensions that leads to an acceptable Megaverse with our universe embedded within it.

The general conclusions of our development are:

Our universe is a 32 dimension Complex Quaternion Space. It is equivalent to QUeST, which has a 3 + 1 Complex Quaternion Dimensions Space-Time and UST Internal Symmetries.

Upon restriction of QUeST to real-valued coordinates, QUeST Becomes the Unified SuperStandard Theory (UST).

The Megaverse is a 32 dimension Complex Octonion Space. It is equivalent to MOST, which has a 7 + 1 dimensions Complex Quaternion Space-Time and Internal Symmetries.

The 7 + 1 dimensions Complex Quaternion Space-Time of the MOST Megaverse contains 3 + 1 dimensions Complex Quaternion QUeST Universes' Space-Times.

QUeST → real-valued coordinates Unified SuperStandard Theory

We show that the Unified SuperStandard Theory is an exact consequence of Quaternion and Octonion space. *The generality of these results, and the author's previous derivation of the exact value of the Fine Structure Constant α, and approximate derivations of the Weak and Strong coupling constants (shown later), shows that the Internal Symmetries of the Standard Model, and our Unified SuperStandard Theory, are now understandable.*

Thus particle physics can be explained by Quaternion and Octonion space characteristics. We show this in some detail in the following chapters.

1. Extension of the Unified SuperStandard Theory (UST) to Higher Dimensions

1.1 Suggestions of a Deeper Basis for UST

The UST provides a complete theory of elementary particles and Gravitation in 3 + 1 dimensions (real-valued coordinates). It is presented in detail in Blaha (2020c). In the study of this theory extending back 7+ years to Blaha (2012a) and earlier books, the author found a close analogy between the subgroups of the Complex Lorentz Group and the factors of the Standard Model internal symmetry: SU(2)⊗U(1)⊗SU(3).

This seeming coincidence raised the question that the analogy was based on a deeper relation embodied in a larger space. This larger space would have a subspace devoted to space-time and a subspace for internal symmetry groups. The subspaces would be orthogonal.

Pursuing this concept lead the author, in the fall, 2019, to develop a formulation of larger spaces based on hypercomplex coordinates: quaternions and octonions. This development reached fruition in Blaha (2020a), (2020b) and (2020c)..

1.2 Hypercomplex Numbers

The key concept in the search for a deeper basis for UST was the following progression of coordinate system formulations:

A. The Standard Model was based on real-valued coordinates.

B. UST is based on real-valued coordinates made complex:

B1. Real-valued coordinates were replaced by slightly complex coordinates with an imaginary part consisting of a massless second quantized vector field that gave the usual perturbation theory results at low energy and \eliminated divergences at very high energy

B2. Quarks and Strong Interaction gauge fields were truly complex (with certain restrictions) due to Complex Lorentz Group boosts.

C. Having progressed from real-valued coordinates to complex-valued coordinates in parts A and B, we considered hypercomplex-valued coordinates and developed 32 complex quaternion coordinate systems and 32 dimension complex octonion coordinate systems.[1] We suggested complex quaternion coordinates were appropriate for our universe, and complex octonion coordinate systems were appropriate for a Megaverse (Multiverse) should our universe reside in it. The author found that the complexity of quaternion space and octonion space was important for the analogy with complex space-time of UST. This is discussed in Blaha (2020c) and later in this book.

D. The choice of 32 dimension coordinate systems was motivated by the perceived need to have a complex quaternion space-time for universes and a complex octonion space-time for the Megaverse. A benefit of this approach was that it forced layers of particles to exist in the resulting theories—consistent with the layers found in the real-valued coordinates UST.

The result of this chain of progression was a complex quaternion space-time theory (called QUeST) that directly becomes UST upon restriction to real-valued coordinates. Further, upon restriction of the complex octonion space-time theory (called MOST) to a complex quaternion space-time we obtain QUeST. Thus there are three closely related theories UST, QUeST, and MOST. We discuss details related to this section later.

---

[1] The use of quaternion and octonion number systems is predicated on the nature of S matrix elements, which necessarily use the Dyson-Wick expansion of time ordered products in perturbation theory. S matrix elements require the rules of real numbers or complex numbers or quaternion numbers or octonion numbers for calculations.

1.3 Mathematical Picture Language

As we will see later we need only consider the allocation of dimensions to space-time, and to the various internal symmetry groups in the 32 dimension spaces. The detailed mathematics of quaternions and octonions is not directly relevant for the internal symmetry subspaces. The quaternions and octonion coordinates of the space-times can utilize the algebras of these hypercomplex numbers in calculations.

Since the allocation[2] of dimensions to fundamental representations of groups is of paramount importance to establish the basis of UST we simply used the symbol • to represent a dimension. The many dimensions of the 32 dimension spaces, which take account of the dimensions within a quaternion quartet, and an octonion octet, are thus represented by patterns of •'s. The separation of the dimensions (•'s) into fundamental representations is then easily accomplished.

Thus we have a simple Mathematical Picture Language for the separation that would require complex transformations otherwise as indicated in the footnote below.

1.4 Number Describes Everything

Pythagoras developed a language of •'s later for computation. He called the •'s *psiphi* meaning pebbles in Classical Greek. In this regard he developed a set of diagrams for numbers. The perfect number 10 was represented by

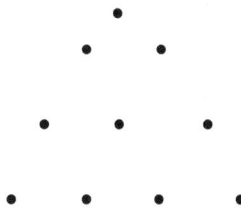

```
        •
      •   •
    •   •   •
  •   •   •   •
```

Figure 1.1. Pythagorean tetrackys, a psiphi representation of 10.

---

[2] The allocation of dimensions is a euphemism for the use of transformations to block diagonalize the 32 dimension spaces.

He considered ten to be the perfect number because it embodies 1, 2, 3, and 4. Remarkably perhaps, the groups U(1), U(2), SU(3), and U(4) are the internal symmetry groups of UST and the Standard Model.

Pythagoras' great achievements and his innate sense of the nature of Reality led to the proposition: Our theories directly reflect its truth.

$$\textbf{The essence of all things is number.} \tag{1.1}$$

From total symmetry of the internal symmetry subspace, and the use of the number of dimensions of fundamental representations of internal symmetry groups, we are led directly to the internal symmetry groups, the fermion spectrum, the vector boson spectrum and Gravitation.

We will see "blocks" of fundamental representations of the groups known from the Standard Model (and UST) appearing in the quaternion and octonion theories:

$$SU(2) \otimes U(1) \otimes SU(3) \otimes SU(2) \otimes U(1) \otimes SU(3) \tag{1.2}$$

where the second set of $SU(2) \otimes U(1) \otimes SU(3)$ factors are symmetries of Dark matter in UST. The total dimension of these factors is 10, the Pythagorean perfect number, which can be viewed as an interesting coincidence. It suggests the Pythagorean summation of Reality should be:

$$\textbf{Number is the Essence of Eternity} \tag{1.3}$$

since one can hope the theories, UST, QUeST, and MOST, if correct, are eternal being all based on Number.

1.4 Axioms for the Unified SuperStandard Theories

We list a set of Axioms for the combined Unified SuperStandard Theories for our 3 + 1 dimension universe, for the 3 + 1 complex quaternion QUeST universe, and for the 7 + 1 octonion MOST Megaverse.

AXIOMS

1. A biquaternion space is the basic space-time of our universe. Bioctonion space is the basic space-time of the Megaverse.

2. Physical processes can execute in parallel.

3. Matter and energy are particulate.

4. Space--times are locally Lorentzian.

5. All calculations are finite.

6. Particle theory can be defined in any curved space-time.

7. Each particle has a wave function determined by a functional inner product defining the particle state. The functionals form a set without a distance measure.

1.5 General Implications of the Axioms

In this section we describe some of the implications of each of the axioms.

1. Biquaternion space is the basic space-time of our universe. Bioctonion space is the basic space-time of the Megaverse.

The factorization into a space-time and an internal symmetry space must be a form of spontaneous symmetry breaking of yet unknown origin. It appears to be related to a breakdown of the vacuum.

2. Physical processes can execute in parallel.

Physical processes are known to be able to execute in parallel at any distance of separation. As Fant has shown parallel execution requires a minimal number of dimensions: 4. Consequently the dimension of space-time must be 4 or greater. The biquaternion space-time of QUeST is 4-dimensional allowing parallel process execution.

The bioctonion space-time of MOST is 8-dimensional and also allows parallel process execution. The choice of eight dimensions is natural since it allows 4-dimensional universes within it. It also has a form that allows a clean formulation. Lastly, as will be seen later, it conforms to the pattern of interplay between Lorentz symmetry and internal symmetry found in the Unified SuperStandard Theory. This axiom leads to a view of the origin of the dimensions.

3. Matter and energy are particulate.

The most direct method of specifying a theory of matter and energy is through the Use of Quantum Field Theory. Thus Quantum Field Theory is implied.

4. Complex Space-times are locally Lorentzian.

A locally complex Lorentzian space-time leads to Complex General Relativity. In flat space-time Complex General Relativity becomes Complex Lorentz group. (In point of fact the Complex Poincaré group follows.)

5. All calculations are finite.

Given the need for Quantum Field Theory it becomes necessary to find a formulation that yields finite values for calculations in perturbation theory. The only approach that eliminates high energy divergences, and yet preserves the results found in perturbation theory calculations that agree with (primarily QED) experiments, is Two-Tier Quantum Field Theory. This is discussed in detail in earlier books starting in 2002. Thus only our Two-Tier formalism satisfies this axiom.

6. Particle theory can be defined in any curved space-time.

In the 1970s we developed a formalism that allows the definition of particle states in any space-time in such a way that its physical content is preserved when transformed to any coordinate system.[3] This PseudoQuantum Quantum Field Theory satisfies this axiom.

7. Each particle has a wave function determined by a functional inner product defining the particle state. The functionals form a set without a distance measure.

This axiom is satisfied by our formulation of quantum functionals in Blaha (2019f) and earlier books. Our formulation eliminates the superficial violation of the Theory of Relativity by "spooky" quantum entangled processes with parts separated by a physically "large" distance.

The seven axioms imply the Unified SuperStandard Theory and its deeper biquaternion and bioctonion hypercomplex formulations.

---

[3] S. Blaha, Il Nuovo Cimento **49A**, 35 (1979).

2. Hypercomplex Coordinate Systems

2.1 Hypercomplex Number Based Higher Dimensions

The Unified SuperStandard Theory was based on real-valued and complex-valued coordinates. This choice enabled us to understand the reason behind the four types of fermions found in nature: neutral fermions, charged fermions, up-type quarks, and down-type quarks. A study of Complex Lorentz group subgroups showed that they were similar to the factors of the Standard Model symmetry group. This similarity motivated this author to consider a space that integrated space-time and internal symmetry dimensions.

In choosing a higher dimension space for a larger theory of elementary particles the use of coordinate systems based on hypercomplex number systems seemed reasonable. The hope was that just as complex coordinates led to a deeper understanding of the internal symmetries of the Standard Model, the use of hypercomplex coordinates might lead to a further understanding of the origin of both space-time and internal symmetries.

The use of quaternion and octonion coordinate systems was motivated by the nature of S matrix elements, which necessarily use the Dyson-Wick expansion of time ordered products in perturbation theory.[4] S matrix element calculations require the arithmetic rules of real numbers or complex numbers or quaternion numbers or octonion numbers for calculations.

The pattern of rising hypercomplexity is:

Real → Complex → Quaternion → Biquaternion → Octonion → Bioctonion

---

[4] Time-ordered products are defined in quaternion perturbation theory in terms of a "time" defined as a quaternion inner product of a chosen direction in quaternion time, and the time quaternion. Quaternion time is thus single-valued. Potential infinities in higher dimension perturbation theory are eliminated using the author's Two Tier formulation of Quantum Field Theory. See Blaha (2005a).

The Unified SuperStandard Theory took particle theory from real-valued coordinates to complex-valued coordinates. Complex quaternion (biquaternion) and complex octonion (bioctonion) extensions took us to QUeST and MOST. Both use larger spaces to unite space-time symmetry and internal symmetry.

Since the requirement of parallel physical processes[5] made the minimal space-time dimension 4 and since the Megaverse must include universes as subspaces, we were led to a 4-dimensional complex quaternion formulation for our universe and an 8-dimensional complex octonion formulation for the Megaverse. They were considered in Blaha (2020a), (2020b) and Blaha (2020c). Generalizations to higher dimensional quaternion and octonion spaces were considered to accommodate the four layers suggested by the Unified SuperStandard Theory.

In Blaha (2020c) we considered these higher dimensional theories and found that the complex quaternion higher dimensional theory QUeST) leads directly to the 3 + 1 dimensional Unified SuperStandard Theory upon restriction of the theory to real-valued coordinates. Similarly the Megaverse MOST leads to a reasonable Megaverse theory upon restriction to a generalization of the Unified SuperStandard Theory in 7 + 1 real space-time dimensions.

2.2 From Complex Coordinates to Hypercomplex Coordinates

In this section we show a natural generalization of coordinate systems to hypercomplex number systems. There are a number of ways to define hypercomplex coordinate systems. We extrapolate from complex coordinates:

$$t = t_1 + it_2$$
$$x = x_1 + ix_2$$
$$y = y_1 + iy_2$$
$$z = z_1 + iz_2$$
$$\dots$$

Note that the real part of each coordinate is the same type as the imaginary part—also a real number.

---

[5] See Blaha (2020c).

2.3 Quaternion Coordinate Systems

Based on that simple fact it seems natural to define a 3 + 1 dimensional "real-valued" space with quaternion coordinates as:

Time quaternion
$$t = (a + ib + jc + kd) \tag{2.1}$$
Spatial quaternions
$$x = (a_x + ib_x + jc_x + kd_x)$$
$$y = (a_y + ib_y + jc_y + kd_y)$$
$$z = (a_z + ib_z + jc_z + kd_z)$$

where a, b, c, d are real-valued numbers. The symbols i, j, and k are fundamental quaternion units. Quaternions are described in Appendix 2-A. A quaternion embodied four coordinates. Thus it is a 4-dimensional entity and we attribute 4 dimensions to each quaternion and 8 dimensions to each complex quaternion.

In the UST development (Blaha (2020c)) we saw that complex coordinates and the Complex Lorentz Group were needed to find the four types (species) of fundamental fermions[6] and the internal symmetry groups. Consequently we will use complex quaternions (biquaternions) to develop QUeST, initially in a 7 + 1 complex quaternion space.

Time Biquaternion
$$t = (a + ib + jc + kd) + I(a' + ib' + jc' + kd') \tag{2.2}$$
Spatial Biquaternions
$$x = (a_x + ib_x + jc_x + kd_x) + I(a'_x + ib_x' + jc_x' + kd_x')$$
$$y = (a_y + ib_y + jc_y + kd_y) + I(a'_y + ib_y' + jc_y' + kd_y')$$
$$z = (a_z + ib_z + jc_z + kd_z) + I(a'_z + ib_z' + jc_z' + kd_z')$$
$$x1 = (a_{x1} + ib_{x1} + jc_{x1} + kd_{x1}) + I(a'_{x1} + ib_{x1}' + jc_{x1}' + kd_{x1}')$$
$$y1 = (a_{y1} + ib_{y1} + jc_{y1} + kd_{y1}) + I(a'_{y1} + ib_{y1}' + jc_{y1}' + kd_{y1}')$$
$$z1 = (a_{z1} + ib_{z1} + jc_{z1} + kd_{z1}) + I(a'_{z1} + ib_{z1}' + jc_{z1}' + kd_{z1}')$$

---

[6] Charged leptons, neutral leptons, up-type quarks, and down-type quarks.

$$w1 = (a_{w1} + ib_{w1} + jc_{w1} + kd_{w1}) + I(a'_{w1} + ib_{w1}' + jc_{w1}' + kd_{w1}')$$

where I is an additional fundamental quaternion unit. Note that the real and imaginary part of each coordinate has the same fundamental quaternion units to permit complex rotations between them.

As we will see we need four iterations of the above set of complex quaternions for the space of QUeST so that it will yield UST upon restriction to real-valued coordinates. Thus QUeST requires a 32 dimension complex quaternion space:

$$t = (a + ib + jc + kd) + I(a' + ib' + jc' + kd') \tag{2.3}$$
$$x = (a_x + ib_x + jc_x + kd_x) + I(a'_x + ib_x' + jc_x' + kd_x')$$
$$y = (a_y + ib_y + jc_y + kd_y) + I(a'_y + ib_y' + jc_y' + kd_y')$$
$$z = (a_z + ib_z + jc_z + kd_z) + I(a'_z + ib_z' + jc_z' + kd_z')$$
$$x1 = (a_{x1} + ib_{x1} + jc_{x1} + kd_{x1}) + I(a'_{x1} + ib_{x1}' + jc_{x1}' + kd_{x1}')$$
$$y1 = (a_{y1} + ib_{y1} + jc_{y1} + kd_{y1}) + I(a'_{y1} + ib_{y1}' + jc_{y1}' + kd_{y1}')$$
$$z1 = (a_{z1} + ib_{z1} + jc_{z1} + kd_{z1}) + I(a'_{z1} + ib_{z1}' + jc_{z1}' + kd_{z1}')$$
$$w1 = (a_{w1} + ib_{w1} + jc_{w1} + kd_{w1}) + I(a'_{w1} + ib_{w1}' + jc_{w1}' + kd_{w1}')$$

$$\cdots$$

$$w4 = (a_{w4} + ib_{w4} + jc_{w4} + kd_{w4}) + I(a'_{w4} + ib_{w4}' + jc_{w4}' + kd_{w4}')$$

Our primary interest in quaternion space is its dimensions and how they can be transformed into a space-time part and an internal symmetry part. Since this purpose is best approached by simply displaying a Mathematical Picture Language representation using Pythagoras' *psiphi* (pebbles) diagrams where each dimension is represented by an • in Fig. 2.1.

The real dimensions[7] of the 7 + 1 space of eq. 2.2 number 64. The real dimensions of the 7 + 1 space of eq. 2.3 number 256. In chapter 3 we will select subsets of dimensions identifying each subset with the fundamental representation of a group.

---

[7] Real-valued coordinates are said to have real dimensions. Complex coordinates have complex dimensions.

Thus the spaces of eqs. 2.2 and 2.3 specify space-time and internal symmetry groups numerically. Then the internal symmetries specify the fermion and vector boson spectrums. Thus the consideration of quaternion spaces, which leads to UST and QUeST, devolves into Number as conjectured by the Pythagorean School.

Time

•••• ••••

Space

•••• ••••

•••• ••••

•••• ••••

•••• ••••

•••• ••••

•••• ••••

•••• ••••

Figure 2.1. Psiphi diagram of the dimensions of 7 + 1 complex quaternion space. Each row represents a complex quaternion. Note the coordinates of eq. 2.2 map directly to dimensions represented by •'s. Each complex quaternion embodies 8 dimensions.

2.4 Octonion Coordinate Systems

The use of octonion coordinate systems is consistent with the calculation of S matrix elements using the Dyson-Wick expansion of time ordered products in perturbation theory.[8] S matrix element calculations require the arithmetic rules of real numbers or complex numbers or quaternion numbers or octonion numbers for calculations. Octonions can be used to define a coordinate system. An octonion has 8

---

[8] Time-ordered products are defined in octonion perturbation theory in terms of a "time" defined as a octonion inner product of a chosen direction in octonion time, and the time octonion. Octonion time is thus single-valued. Potential infinities in higher dimension perturbation theory are eliminated using the author's Two Tier formulation of Quantum Field Theory. See Blaha (2005a).

coordinates and thus it has dimension 8. See Appendix 2-B for a brief description of octonions.

We can define a 3 + 1 dimensional real octonion space as:

Time Octonion

$$t = a + ib + jc + kd + h'a' + i'b' + j'c' + k'd' \qquad (2.4)$$

Spatial Octonions

$$x = a_x + ib_x + jc_x + kd_x + h'a'_x + i'b_x' + j'c_x' + k'd_x'$$
$$y = a_y + ib_y + jc_y + kd_y + h'a'_y + i'b_y' + j'c_y' + k'd_y'$$
$$z = a_z + ib_z + jc_z + kd_z + h'a'_z + i'b_z' + j'c_z' + k'd_z'$$

where a, b, c, d, a', b', c', d' are *real-valued* numbers. The symbols i, j, k, h', i', j' and k' are fundamental octonion units.

In the MOST development (Blaha (2020c)) we saw that complex octonions were needed initially in a 7+ 1 dimensions complex octonion space, which was subsequently expanded four-fold to a 32 dimension complex octonion space. The 7 + 1 dimensions complex octonion space coordinates are:

Time Octonion

$$t = a + ib + jc + kd + h'a' + i'b' + j'c' + k'd' \qquad (2.5)$$

Spatial Octonions

$$x = a_x + ib_x + jc_x + kd_x + h'a'_x + i'b_x' + j'c_x' + k'd_x'$$
$$y = a_y + ib_y + jc_y + kd_y + h'a'_y + i'b_y' + j'c_y' + k'd_y'$$
$$z = a_z + ib_z + jc_z + kd_z + h'a'_z + i'b_z' + j'c_z' + k'd_z'$$
$$x1 = a_{x1} + ib_{x1} + jc_{x1} + kd_{x1} + h'a'_{x1} + i'b_{x1}' + j'c_{x1}' + k'd_{x1}'$$
$$y1 = a_{y1} + ib_{y1} + jc_{y1} + kd_{y1} + h'a'_{y1} + i'b_{y1}' + j'c_{y1}' + k'd_{y1}'$$
$$z1 = a_{z1} + ib_{z1} + jc_{z1} + kd_{z1} + h'a'_{z1} + i'b_{z1}' + j'c_{z1}' + k'd_{z1}'$$
$$w1 = a_{w1} + ib_{w1} + jc_{w1} + kd_{w1} + h'a'_{w1} + i'b_{w1}' + j'c_{w1}' + k'd_{w1}'$$

where all coefficients: a, b, c, d, a', b', c', d', and a_i, b_i, c_i, d_i, a'_i, b'_i, c'_i, d'_i for i = x, y, z, w, x1, y1, z1, w1 are *complex-valued* numbers. Each complex octonion embodies 16 dimensions. See Fig. 2.2 for the psiphi representation of dimensions in eq. 2.5.

As we saw in Blaha (2020c) we need four iterations of the above set of complex octonions for the space of MOST for reasons discussed later. Thus MOST requires a 32 dimension complex octonion space. The number of real dimensions in this space is 512. The coordinates of 32 dimension complex octonion space are:

Time Octonion

$$t = a + ib + jc + kd + h'a' + i'b' + j'c' + k'd' \qquad (2.6)$$

Spatial Octonions

$$x = a_x + ib_x + jc_x + kd_x + h'a'_x + i'b_x' + j'c_x' + k'd_x'$$
$$y = a_y + ib_y + jc_y + kd_y + h'a'_y + i'b_y' + j'c_y' + k'd_y'$$
$$z = a_z + ib_z + jc_z + kd_z + h'a'_z + i'b_z' + j'c_z' + k'd_z'$$
$$x1 = a_{x1} + ib_{x1} + jc_{x1} + kd_{x1} + h'a'_{x1} + i'b_{x1}' + j'c_{x1}' + k'd_{x1}'$$
$$y1 = a_{y1} + ib_{y1} + jc_{y1} + kd_{y1} + h'a'_{y1} + i'b_{y1}' + j'c_{y1}' + k'd_{y1}'$$
$$z1 = a_{z1} + ib_{z1} + jc_{z1} + kd_{z1} + h'a'_{z1} + i'b_{z1}' + j'c_{z1}' + k'd_{z1}'$$
$$w1 = a_{w1} + ib_{w1} + jc_{w1} + kd_{w1} + h'a'_{w1} + i'b_{w1}' + j'c_{w1}' + k'd_{w1}'$$

$$\cdots$$

$$w4 = a_{w4} + ib_{w14} + jc_{w4} + kd_{w4} + h'a'_{w4} + i'b_{w4}' + j'c_{w4}' + k'd_{w4}'$$

where all coefficients: a, b, c, d, a', b', c', d', and a_i, b_i, c_i, d_i, a'_i, b'_i, c'_i, d'_i for i = x, y, z, w, x1, y1, z1, w, ... , w4 are *complex-valued* numbers.

For the Megaverse we will (in chapter 7) select subsets of dimensions identifying each subset with the fundamental representation of a group. *Thus the spaces of eqs. 2.5 and 2.6 specify space-time and internal symmetry groups numerically. Then the internal symmetries specify the fermion and vector boson spectrums. Thus the consideration of octonion spaces leads to MOST, which devolves into Number as conjectured by the Pythagorean School. All is Number!*

Time

• • • • • • • •   • • • • • • • •

7-Space

• • • • • • • •   • • • • • • • •
• • • • • • • •   • • • • • • • •
• • • • • • • •   • • • • • • • •
• • • • • • • •   • • • • • • • •
• • • • • • • •   • • • • • • • •
• • • • • • • •   • • • • • • • •
• • • • • • • •   • • • • • • • •

Figure 2.2. Psiphi diagram of the dimensions of 7 + 1 complex octonion space. Each row represents a complex octonion. Note the coordinates of eq. 2.2 map directly to pairs of dimensions represented by •'s. Each complex octonion embodies 16 real dimensions.

Appendix 2-A. Quaternion Features

Quaternions and octonions are hypercomplex numbers with special properties that make them similar to complex numbers.[9] Quaternions and octonions are both normed division algebras over the reals (*hypercomplex* number systems) with salutary properties for quantitative studies in quantum field theory and perturbation theory. Some of their new features are listed on the cover page.

Quaternions have significant properties that distinguish them:

1 .They are associative.

2. They are one of the two finite dimensional division rings having the real numbers as a proper subring. (The other is octonions—considered in chapter 8.)

3. They are non-commutative. (This is not a roadblock for quantum field theory which is also non-commutative in general.)

These features support the development of physics theories.[10]

2-A.1 Some Basic Quaternion Features

A quaternion is a 4-tuple of real numbers. A complex quaternion is a 4-tuple of complex numbers:

$$x = a + bi + jc + kd \quad = a + \mathbf{v} \tag{2-A.1}$$

---

[9] Much of this appendix appears in Blaha (2020a) and (2020b).

[10] There is an extensive literature on quaternions starting with the original work of Hamilton. Some recent, relevant papers are: S. L. Adler, "Generalized Quantum Dynamics", IASSNS –HEP-93/32 (1993); S. De Leo, arXiv:hep-th/9506179 (1995); Rolf Dahm, arXiv:hep-th/9601207 (1996); S. De Leo, arXiv:hep-th/9508011 (1995); S. L. Adler, arXiv:hep-th/9607008 (1996) and references therein.

where a, b, c, d are real or complex numbers, and **v** is a 3-vector. The symbols i, j, and k are fundamental quaternion units. A quaternion norm is defined by

$$\|\mathbf{x}\| = \text{sqrt}(aa^* + bb^* + cc^* + dd^*) \qquad (2\text{-A}.2)$$

and the norm of **v** is

$$\|\mathbf{v}\| = \text{sqrt}(bb^* + cc^* + dd^*) \qquad (2\text{-A}.3)$$

An important identity is

$$e^x = e^a (\cos(\|\mathbf{v}\|) + \mathbf{v}/\|\mathbf{v}\| \sin(\|\mathbf{v}\|))s \qquad (2\text{-A}.4)$$

.It is used to define boosts in quaternion space.

Appendix 2-B. Some Octonion Features

Octonions have significant properties that enable them to be used in a quantum field theory development:

1. An octonion is an 8-tuple of real numbers. A complex octonion is an 8-tuple of complex numbers.
2. They are nonassociative.
3. They are one of the two finite dimensional division rings having the real numbers as a proper subring. (The other is quaternions—considered in chapter 5.)
4. They are non-commutative. (This is not a roadblock for quantum field theory which is also non-commutative in general.)

These features support the development of physics theories.
We can represent a complex octonion (bioctonion) b as

$$b = b_{real} + Ib_{imaginary}$$

where b_{real} and $b_{imaginary}$ are real-valued octonions. We can also represent a complex octonion as an octonion with complex coordinates as in eq. 2.5.

3. Quaternion Unified SuperStandard Theory (QUeST)

In Blaha (2020a), (2020b) and (2020c) we developed QUeST. This chapter expands on those books.

3.1 Motivation and Procedure

Our initial goal was to create a larger dimension space within which we can derive our space-time and the Unified SuperStandard Theory in such a way as to understand the analogy between Lorentz subgroups and Standard Model internal symmetry groups that we found in Blaha (2018f) and earlier books. We expected the development of a deeper form of the Unified SuperStandard Theory (UST) to lead to refinements.

We found some significant new features—most importantly U(2) groups that transformed between normal and Dark matter. These groups must have extremely large mass vector bosons and/or extremely weak coupling constants to be in accord with experimental observations.

There are two possible procedures to follow in developing the deeper basis:

1. One can develop the Quantum Mechanics and Quantum Field Theory in a quaternion space and then extract the dynamics, fermion spectrum, gauge fields, and so on of our familiar space-time.

2. One can define a quaternion space, and then using its coordinates, directly extract the space-time, internal symmetries, fermion spectrum, gauge field spectrum and dynamics. Quaternion algebra may not be used initially. A quaternion space serves to give shape to the range of possible coordinate systems.

We chose the latter approach as it led directly to the Unified SuperStandard Theory in 3 + 1 real-valued dimensions.[11]

In developing the deeper space, upon which we built, we took guidance from the derivation of the Unified SuperStandard Theory. That theory assumes a complex 4-dimensional space-time upon which Complex General Relativity is constructed. It then proceeds to complex flat space-time and Complex Relativity. Next it restricts the complex coordinates of the theory to real-valued coordinates (excepting quarks and color gluons that remain complex as described in Blaha (2020c)). The four types of fermions followed from Complex Lorentz boosts and the internal symmetries reflected the subgroups of the Complex Lorentz Group.

3.2 Definition of A Complex Quaternion (Biquaternion) Space

Following the stated procedure we defined a complex quaternion (biquaternion) space with one "time" biquaternion and seven "spatial" biquaternions believing the 3+1 space-time of our experience is a consequence of this deeper level. Eq. 2. 2 has the 7 + 1 complex quaternion coordinate system.

We do not use the algebra of quaternions but simply treat the quaternion coordinates as coordinates in a space. Our only need was to determine internal symmetries based on the dimensions of fundamental group representations: 4 real dimensions for SU(2)⊗U(1); 6 real dimensions for SU(3); and 8 real dimensions for U(4).

Fig 2.1 symbolically depicts the dimensions of the 7 + 1 complex quaternion space. This space has 64 real dimensions (32 complex dimensions.).

3.2.1 Dimension Functionals

The dimension "array" in Fig. 2.1 can be represented in terms of *Dimension Functionals*. We can define a dimension row functional D_i that takes a column dimension as its argument and generate the dimension array $[d_{ij}]$ of Fig. 2.1 with

$$d_{ij} = D_i(d_j) \qquad (3.1)$$

---

[11] This was an unexpected, but welcome, surprise for the author.

where $i = 1, 2, \ldots, 8$ and $j = 1, 2, ---, 8$. A dimension column functional may be defined that enables a dimension array to be generated by a functional E_j from a single initial dimension:

$$d_j = E_j(d) \tag{3.2}$$

Combining eqs. 3.1 and 3.2 we may generate a dimension array with

$$d_{ij} = D_i(E_j(d)) \tag{3.3}$$

from a single dimension d. We can envision a universe created from a single dimension using this formalism and a suitable dynamics.

Dimension functionals can also be used in the case of octonion coordinates as seen later.

This formulation illustrates the use of dimensions as functionals. We will discuss functionals later where we map dimensions to particle functionals.

3.3 Biquaternion Lorentz Group

Our definition of time and space biquaternion coordinates purposefully resembles those of our real space-time. One might ask why should there be a Lorentz-like group for biquaternion space. The only apparent reason is the need for a special speed c in our space-time that enables one to boost from a rest frame of a mass m particle with energy m to a moving frame of greater energy.[12] Without c the group of the above coordinates would presumably be the biquaternion U(8) group. In this group, transformations preserve the norm of a state so that a "boost-like" transformation does not exist.

The biquaternion Lorentz transformations do have a unique speed c (the speed of light) and specify a unique rest frame for any particle—both sublight particles and

---

[12] A similar consideration applies to tachyons.

tachyon particles.[13] Thus we select the Biquaternion Lorentz group SU(7, 1) for 7 + 1 dimension biquaternion space.

Flat space biquaternion Special Relativity generalizes directly to a biquaternion General Relativity which may be constructed directly (mindful of quaternion non-commutativity).

The flat space biquaternion Lorentz group transformations have constant biquaternion matrix elements that are analogous to those of the Lorenz group.. (See Blaha (2020c) for Lorentz group boosts.)

Motions in complex quaternion space are of four types: motions where the norm of the velocity is real-valued and less than c; motions where the norm of the velocity is real-valued and greater than c; motions where the norm of the velocity is complex-valued and whose absolute value is less than c; and motions where the norm of the velocity is complex-valued and whose absolute value is greater than c,

Boosts generating these cases lead to the separation of fermions into four species for the Unified SuperStandard Theory and its complex quaternion basis QUeST (as well the complex octonion Megaverse MOST.)

3.4 Extracting the Symmetries and Particle Spectra

As stated earlier in section 3.2 we will directly describe the symmetry structure implied by the form of the biquaternion coordinate system while mindful of section 3.3.

The Unified SuperStandard Theory developed the group structure from which the particle species were derived from a subset of Lorentz boost transformations. It used boosts with complex exponentiation similar to quaternion exponentiation in eq. 2-A.4. Complex boosts mapped a system at rest to a system in motion with a real energy and complex 3-momenta in general. Biquaternion boosts play a similar role.

The 4-dimensional subspace for the Unified SuperStandard Theory complex coordinate boosts' was

---

[13] Tachyons can be transformed by a Complex Lorentz transformation to and from a rest frame. See Blaha (2018f).

Time

•

Space

• •

• •

• •

Figure 3.1. Four-Dimensional subspace for Unified SuperStandard derivation of particle spectra with coordinates represented by psiphi • 's. Note the time coordinate is real-valued.

    Following the same line of reasoning we now specify a biquaternion subspace analogously restricted to that of Fig. 3.1 to define the relevant set of coordinates for determining particle symmetries and spectra. Note time is a "real" quaternion in this subspace. The spatial coordinates consist of complex quaternions.

Time

• • • •

Space

• • • •   • • • •

• • • •   • • • •

• • • •   • • • •

• • • •   • • • •

• • • •   • • • •

• • • •   • • • •

• • • •   • • • •

Figure 3.2. 8-dimensional subspace for determining internal symmetries and particle spectra.

We expect its 30 complex dimensions (60 real dimensions) will split into a 4-dimensional complex coordinate space[14] which will support Complex Lorentz transformations, and the remaining 26 complex coordinates will represent internal symmetry space. Internal symmetry space has an initial U(26) group before breakdown.

The mechanism for this symmetry breakdown may involve vacuum energy effects in the biquaternion universe.

The remaining four time dimensions (coordinates) shown in Fig. 2.1 will be shown to correspond to a U(2) group that maps between normal fermions and Dark fermions on a one-to-one basis. Experiment indicates this group's interactions must be extraordinarily weak.

In the next chapter we analyze the symmetries of the 26 complex dimensional subspace, and discuss the U(2) group in more detail.

---

[14] The 7 + 1 dimension biquaternion space yields a 3 + 1-dimensional complex coordinate space which then becomes the 3 + 1 dimensional real-valued space of our experience.

4. Internal Symmetries of the Quaternion Unified SuperStandard Theory (QUeST)

The coordinate space picture of QUeST described in chapter 3 enables us to simply find the internal symmetries and particle spectra of QUeST. They will turn out to be those of the Unified SuperStandard Theory.

4.1 QUeST Space

The QUeST coordinate subspace for the determination of internal symmetries is depicted in Fig. 4.1..It is based on the discussion of Fig. 3.2. The complex space-time 4-vector is separated from the internal symmetry coordinates in Fig. 4.1.

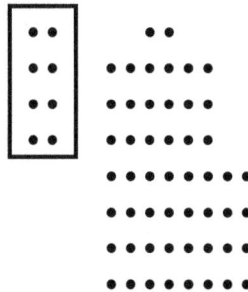

Figure 4.1. Symmetry determination subspace. Eight real-valued space-time dimensions (giving a complex 4-vector) are separated from the 52 real-valued dimensions for internal symmetries. The internal symmetry dimensions are split into subsets for each of the internal symmetries.

The internal symmetry coordinates number 52 real coordinate dimensions or 26 complex coordinate dimensions. These coordinates serve as coordinates of the fundamental representations of each of the factors of the total internal symmetry.

4.2 One Layer QUeST

The 26 complex coordinates form a U(26) internal symmetry space. This space undergoes a breakdown. The resulting factorized form of the internal symmetries[15] is given by

$$SU(2)\otimes U(1)\otimes SU(3)\otimes SU(2)\otimes U(1)\otimes SU(3)\otimes U(4)^4 \qquad (4.1)$$

The 7 + 1 dimension QUeST leads to a one layer form of the Unified SuperStandard Theory. In this theory there is one Layer group with a singlet representation.

When we generalize 7 + 1 QUeST to 32 complex quaternion dimensions we will have a QUeST with four layers that corresponds to the Unified SuperStandard Theory (UST) in 3 + 1 dimensions with the addition of a [U(2)]^4 factor.

The one layer QUeST is represented by Fig. 4.2. The seemingly duplicate factors in eq. 4.1 of

$$SU(2)\otimes U(1)\otimes SU(3) \ \otimes SU(2)\otimes U(1)\otimes SU(3) \qquad (4.2)$$

are for the "normal" and the Dark sectors respectively as shown in Fig. 4.2.[16] Fig. 4.2 reflects a map of the dimensions (coordinates) of Fig. 4.1 into representations of internal symmetry groups.

4.2.1 The U(2) Rotation Group Between Normal and Dark Sectors

The omitted four real-valued dimensions (coordinates) in Fig. 4.1 (compared to the complete set of dimensions in Fig. 2.1) are for the fundamental representation of a U(2) group. It appears that the only role this group could play is to transform between the Normal and Dark sectors. Thus we call it the *Dark Group*. It transforms each normal

---

[15] Eq. 4.1 will be shown later to correspond to *one layer* of UST. UST has four layers, which are described later in this chapter.

[16] Simple counting of fundamental representation dimensions shows this to be true:: 2 + 3 + 2 + 3 = 10 respectively. The set of 10 complex coordinates support transformations to a factorized block-diagonal form. The 10 complex coordinates can be transformed into fundamental representations of the above factor group product.

fermion to its Dark counterpart and *vice versa* (as well as rotations to fermions intermediate between normal and Dark.)

These observations are based on a realization that the Generation Group transforms among the four generations of the layer, and that the Layer group transforms among the four layers of fermions. Consequently the only remaining role the Dark $U(2)$ group can play is between the normal and Dark fermion spectrums. Together these three types of groups "unite" the complete fermion spectrum. Fig. 4.3 shows the roles of the Generation, Layer and Dark groups for the fundamental fermion spectrum.

Experiment suggests the Dark $U(2)$ group is necessarily broken with large mass vector bosons and/or with ultraweak interactions. Fig. 4.2 displays the one layer internal symmetry groups and complex space-time dimensions.

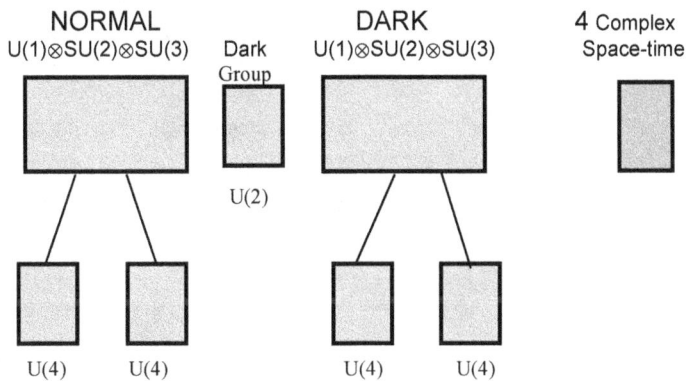

Figure 4.2. Schematic of the internal symmetry groups of eq. 6.1 plus 4 complex dimensions space-time. The two large blocks are each 5 dimension complex coordinate representations of $SU(2) \otimes U(1) \otimes SU(3)$. The $U(2)$ group supports transformations (rotations) between normal and Dark matter.

The Fermion Periodic Table

Figure 4.3. Fermion particle spectrum and partial example of pattern of mass mixing of the Generation, Layer, and Dark grroups. Unshaded parts are the known fermions with an additional, as yet not found, 4th generation shown. The lines on the left side (only shown for one layer) display the Generation mixing

within each layer's species. The Generation mixing applies within each layer using a separate Generation group for each layer. The lines on the right side show Layer group mixing with the mixing amongst all four layers for each of the four generations individually. There are four Layer groups. The Dark groups mixing between normal and Dark fermions are shown in the center as horizontal lines. There are 256 fundamental fermions counting quarks as triplets.

The Dark U(2) group transforms between normal particles and Dark particles. Thus a normal electron is mapped to the equivalent Dark electron, a normal neutrino is mapped to the equivalent Dark neutrino, a normal quark is mapped to the equivalent Dark quark, and so on for all fermions. Similarly normal vector bosons are mapped to Dark vector bosons and normal Higgs particles are mapped to Dark Higgs particles.

Examples of these mappings are:

$$\psi_d = U^{-1}\psi_n$$
$$A^\mu_d = U^{-1}A^\mu_n U$$
$$\psi_n = U\psi_d$$
$$A^\mu_n = UA^\mu_d U^{-1}$$

where d indicates Dark, n indicates normal, ψ is a fermion field, and A^μ is a vector boson field.

The combination of Generation, Layer and Dark groups interconnects, in principle, all parts of the fermion spectrum shown in Fig. 4.3. Thus "isolated" fermions do not exist in UST or QUeST.

4.2.2 Generation and Layer Groups

The lower U(4) groups in Fig. 4.2 are the Generation and Layer number groups. One pair of each number group is for each of the two $U(1)\otimes SU(2)\otimes SU(3)$ factors above. Layer and Generation groups are part of the Unified SuperStandard Theory. See chapters 44 and 45 of Blaha (2020c) for detailed discussions.

4.3 The Breakdown of the Symmetry of the 20 Real Coordinate Group SU(10)

The 20 real coordinate subspace of Fig. 4.1 map to fundamental representations of

$$SU(2) \otimes U(1) \otimes SU(3) \otimes SU(2) \otimes U(1) \otimes SU(3) \otimes U(4)^4 \qquad (4.3)$$

which clearly can be transformed to a block diagonal form in a 26 dimensional group factors representation.[17]

However it is instructive to consider the generation of the component factors using a Lorentz boost-like framework. Separating the internal symmetry part[18] of Fig. 4.1 into six sets of Lorentz-like coordinates with a real energy and complex momenta we obtain Fig. 4.4.

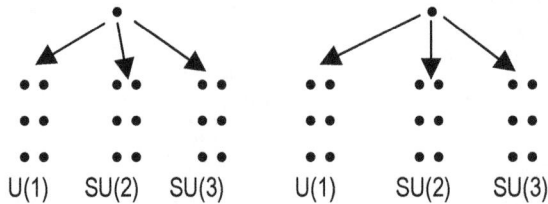

Figure 4.4. Each of the two real-valued time coordinates is linked to three complex spatial coordinates. Lorentz transformations applied to each 4-vector so constructed yields a factor of eq. 4.3. The factors so generated commute to give eq. 4.3 with commuting subgroup factors listed in the figure.

Each of the six sets of Lorentz-like coordinates have an embedded subgroup (listed in Fig. 4.4). These subgroups commute. Then taking each of the six different subgroups

---

[17] The separation of the 20 coordinate subspace into the subgroup factors representation can be implemented as group transformations and definitions using standard group theoretic methods. A more formal method for extracting the subgroup content of representations uses a symmetric group analysis of U(n) representation characters. See S. Blaha, J. Math. Phys. **10**, 2156 (1969) for a detailed discussion of this approach.

[18] Excluding the $U(4)^4$ and $U(2)$ parts.

from the six subsets, that we can form, enables us to form the direct product in eq. 4.2. Thus the analogy between the Standard Model internal symmetries and Complex Lorentz subgroups is validated by the above construction.

4.4 Map to Unified SuperStandard Theory Groups

The map of the four complex coordinates to Complex space-time is direct. Complex Lorentz group transformations relate coordinate systems.

The complete set of internal symmetries (excepting the Dark group) for one layer is given by eq. 4.3. The map follows from the observations:

1. The factors of eq. 4.3 can be separated into separate factors for the normal and Dark sectors.

2. Since the set of normal and Dark fermions are split into four species[19] we can unambiguously associate the factors of eq. 4.3 with its factors.

3. The SU(3) factors have $\underline{3}$ representations which we can associate with the up-quark and down-quark, normal and Dark species. Thus the SU(3) subgroups are normal and Dark Strong interaction subgroups.

4. The SU(2)⊗U(1) factors map to ElectroWeak interactions for the normal and Dark sectors.

5. The four U(4) factors map to the Generation and Layer number groups for the normal and for the Dark sectors.

Thus we have a map to the interactions of the *one layer* Unified SuperStandard Theory. The Dark U(2) group is added to the layer. Although we consider only one layer, the considerations apply to all four layers.

---

[19] Appendix 5-B shows QUeST has four species (types) of fermions.

4.5 Four Layer QUeST Maps Exactly to UST Plus Dark U(2)^4

The QUeST figures above do not display the Internal Symmetry, Generation and Layer groups of all four layers of the Unified SuperStandard Theory. The groups of the layers are not the same. Each layer has its own set of groups;

$$SU(2)\otimes U(1)\otimes SU(3)\otimes SU(2)\otimes U(1)\otimes SU(3)\otimes U(4)^4 \qquad \text{See Fig. 4.2}$$
and
$$U(2) \qquad \text{Section 4.2.1}$$

The overall internal symmetry is the internal symmetry group of the Unified SuperStandard Theory augmented by U(2)^4 giving

$$[SU(2)\otimes U(1)\otimes SU(3)\otimes SU(2)\otimes U(1)\otimes SU(3)\otimes U(4)^4\otimes U(2)]^4 \qquad (4.4)$$

The one layer theory is described by 7 + 1 dimension complex quaternion QUeST. The four layer theory is described by a 32 dimension complex quaternion QUeST. Thus it consists of four "copies" of the coordinates:

Figure 4.5. The 32 complex quaternion dimensions QUeST schematic.

which yield four duplicates of the internal symmetry schematic in Fig. 4.2. Note: there are four U(2) rotation groups – one for each layer,

4.6 Justification for a Four Layer QUeST

There is good reason for QUeST to have four layers embodied in 32 dimension complex quaternion space. If one considers the content of the layer displayed in Fig. 4.2 one sees a 4 dimension complex coordinates block for space-time. To create a 4 dimension complex quaternion coordinates space-time, one needs four layers of the form of Fig. 4.2. *The combination of the four 4-dimension complex coordinates is a complex quaternion dimensions space-time.* Thus the choice of four layer QUeST gives us a 4-dimension complex quaternion space-time AND enables QUeST to map directly to UST with its four layers if one limits the quaternion coordinates to the real-valued coordinates within them.[20]

We conclude four layer QUeST is needed to have a 4 dimension complex quaternion space-time.

---

[20] The Layer groups of UST enable mixing between the layers of fermions as shown in Fig. 4.3.

Appendix 4-A. Gauge Groups Based on Particle Numbers

In this Appendix[21] we show the origin of the Generation and Layer groups in particle number operators since they are not well known. See Blaha (2020c) for more details. Particle interactions followed directly in the Unified SuperStandard Theory by analogy with Complex General Relativity subgroups yielding

$$SU(2) \otimes U(1) \otimes SU(3) \otimes SU(2) \otimes U(1) \otimes SU(3) \qquad (4\text{-}A.1)$$

where the latter three factors are the Dark interactions.

They have a SU(10) covering group that contains this direct product of groups. The groups in eq. 4-A.1 are particle interaction groups in the Unified SuperStandard Theory.

Unlike other attempts to develop a formulation of the Standard Model (or generalizations) the Unified SuperStandard Theory was originally directly based on a theory foundation consisting of Complex General Relativity and Quantum Field Theory.

To those who might prefer to base a theory on real General Relativity we note that proofs in Quantum Field Theory *require* the Complex Lorentz Group.[22] Thus the Complex Lorentz group is unavoidable for a properly (and rigorously) formulated Quantum Field Theory. Since the formulation of the Complex Lorentz Group in flat space-time can only be as the limit of Complex General Relativity, the choice of a foundation of Complex General Relativity is required.

---

[21] This appendix is an extract from Blaha (2020c) for the readers convenience.
[22] Streater (2000).

Since particles are countable, and thus have discrete particle numbers, Quantum Field Theory brings particle numbers, and particle number laws such as particle conservation laws, into consideration.

Blaha (2019e) and earlier books showed that Complex Lorentz boosts generate four types of fermion particles that we call *particle species*. We map these four species to charged leptons (such as electrons), neutral leptons (such as neutrinos), up-type quarks (such as the u quark), and down-type quarks (such as the d quark).

4-A.1 Basis of the Generation Group

We define two particle number operators for normal up-quark particles and down-quark particles, B_{uq} and B_{dq}. Similarly we define two particle number operators for normal species "e" (electron) particles and species "v" particles, B_e and B_v. Similarly we define Dark matter equivalents:[23] B_{De}, B_{Dv}, B_{Duq}, and B_{Ddq}.

In the absence of interactions these fermion particle number operators are conserved. Each set are "diagonal" operators within a U(4) group. Thus we have a normal U(4) Generation Group and a Dark U(4) Generation group.

On this basis we find there are four generations of each species in the normal and in the Dark matter sectors. One generation of normal fermions with large masses has not as yet been found.

The gauge vector bosons of the Generation Group also have large masses. If the conservation of the fermion particle numbers is broken then we view it as a consequence of Generation Group symmetry breaking.

4-A.2 Basis of the Layer Group

The set of particle number operators can be further refined if we take account of the fourfold fermion generations. To further refine the set of particle number operators we temporarily neglect all interactions that would violate conservation laws for the set.

---

[23] By analogy, we assume that there are four species of Dark matter: charged Dark leptons, neutral Dark leptons, Dark up-type quarks, and Dark down-type quarks. Thus we are led to the Dark particle numbers: Dark Baryon Numbers, and Dark Lepton Numbers shown above.

We therefore subdivide the above particle number set into four particle numbers per generation. For the ith generation we define

L_{ie} – The "e" species particle number for the ith generation
L_{iv} – The v species particle number for the ith generation
L_{iuq} – The up-quark species particle number for the ith generation
L_{idq} – The down-quark species particle number for the ith generation

L_{iDe} – The Dark "e" species particle number for the ith generation
L_{iDv} – The Dark v species particle number for the ith generation
L_{iDuq} – The Dark up-quark species particle number for the ith generation
L_{iDdq} – Dark down-quark species particle number for the ith generation

for each generation $i = 1, 2, 3, 4$. Individual fermions have positive $L_{ia} = +1$ values and anti-fermions have negative $L_{ia} = -1$ values for species $a = 1, 2, 3, 4$ (with the three color subspecies of quarks treated as part of one species.)

At this point we have four particle number operators for each generation. We define a group framework for each set of particle numbers. The simplest way is to assume that each generation consists of four layers with the particles in each generation in a U(4) fundamental representation.[24] Then each generation has a U(4) Layer group with the generation's four number operators (above) as its diagonal operators. We call this group the Layer Group of the ith generation L_{ia}. With four generations we obtain four U(4) Layer groups for normal matter. In addition there are four U(4) Dark Layer groups.

The consequence of this expansion of particle numbers and groups is that the set of fermions increases fourfold. We now have four layers, with each having four generations, Experimentally, we know of three generations of fermions—the lowest generations of the lowest level. The remaining generation and three levels of fermions are of much higher mass and yet to be found.

---

[24] See Fig. 2.3 for a depiction of the "splitting" of fermions: first into generations, then into layers.

See Blaha (2019g) and (2018e) for a detailed discussion of the Layer Groups. We note in passing that the symmetries of these number operators are badly broken. Yet the underlying group structure remains.

5. QUeST Universes

QUeST is based on a 32 dimension complex quaternion space. It produces a 3 + 1 dimension complex quaternion space-time and the set of internal symmetries of the UST, to which $U(2)^4$ is added.

The similarities of QUeST and UST include the same four layer pattern; an identical normal and Dark fermion spectrum; and the same internal symmetries and vector bosons with QUeST adding the $U(2)^4$ Dark group vector bosons. See Appendix 5-A for the fermion and vector boson spectrums.

UST has additional detail that is consistent with the real-valued coordinates limit of QUeST. See Blaha (2020c).

Upon restriction of QUeST to real-valued coordinates, QUeST Becomes the Unified SuperStandard Theory.

5.1 Are All Universes the Same?

Assuming that there are other universes beyond our own universe, the question of the Physics of other universes arises. If all universes are 32 dimension complex quaternion universes based on QUeST, then it appears their gross features would be similar. As we showed in Blaha (2020c), and in earlier books, coupling constants appear to be determined by vacuum polarization eigenvalue conditions.

The space-time, Internal Symmetries, fermion spectrum, and the vector boson spectrum are determined in QUeST. Thus the only remaining physical parameters needed are particle masses, and particle mixing. These parameters appear to be handled by vacuum considerations, which should be similar in all sufficiently large universes.

We conclude that all universes are the same based on these considerations. When we consider a complex octonion space, that we call the Megaverse, we will see that it is possible that 6 different varieties of universe may exist within the Megaverse.

Appendix 5-A. Unified SuperStandard Theory (UST) in 3 + 1 Real Dimensions

This Appendix presents a summary of the main consequences of our Unified SuperStandard Theory in 3 + 1 real dimensions. We see that it is the limit of the Quaternion Unified SuperStandard Theory (QUeST) in 32 Complex Quaternion Dimensions when it is restricted to 3 + 1 real-valued space-time coordinates.

The internal symmetries (interactions) of the theory, the vector boson spectrum, and the fermion spectrum are presented. QUeST predicts all these results in 32 dimension complex quaternion space. ***Thus QUeST is the ground of the Unified SuperStandard Theory in 3 + 1 real dimensions. UST was derived previously based on real-valued 3 + 1 dimensional coordinates, Complex General Relativity and Quantum Field Theory.***

5-A.1 Axioms and Derivation of the Unified SuperStandard Theory in 3 + 1 real dimensions

The axioms and derivation of the Unified SuperStandard Theory are presented in Blaha (2020c).

5-A.2 Unified SuperStandard Theory Interactions

The vector boson symmetries[25] in the Unified SuperStandard Theory were found to be

$$[U(1)\otimes SU(2)\otimes SU(3)\otimes U(1)\otimes SU(2)\otimes SU(3)\otimes U(4)^4]^4 \otimes U(4) \qquad (5\text{-}A.1)$$

---

[25] Most of these symmetries are broken symmetries.

QUeST causes us to add a new $U(2)^4$ Dark groups factor. The symmetries are of similar form in each of the four layers. In addition the symmetries of the "normal" sector and the Dark sector are duplicates.

QUeST has the same internal symmetries and adds an additional $U(2)$ symmetry for each of the four layers. Thus $U(2)^4$. This symmetry maps each normal fermion to its corresponding Dark equivalent, and vice versa, layer by layer. We discuss this symmetry in more detail later.

The symmetries of the normal sector are

$$[U(1) \otimes SU(2) \otimes SU(3) \otimes U(4)^2]^4 \otimes U(4) \qquad (5\text{-}A.2)$$

where the lone $U(4)$ is that of the Species group described later.

The symmetries of the Dark sector are similar (but different0

$$[U(1) \otimes SU(2) \otimes SU(3) \otimes U(4)^2]^4 \otimes U(4) \qquad (5\text{-}A.3)$$

where the lone $U(4)$ is again that of the Species group.

Each of the four layers of fermions experiences its own set of interactions. They must be different since inter-layer interactions are not seen. Similarly the normal and Dark sectors of each layer must also have different sets of interactions. The $U(2)$ groups added by QUeST must also yield ultra-weak interactions since interactions between normal and Dark matter are not seen.

The Layer groups act to interconnect the four layers for the normal and Dark layers separately. The $U(2)$ groups serve to connect the normal and Dark sectors (in principle). Thus the symmetries result in an interconnected spectrum of fermions.

5-A.3 Unified SuperStandard Theory Vector Bosons

Fig. 5-A.1 displays the Unified SuperStandard vector bosons except for the Species group which all particles experience, and except for the $U(2)$ groups added by QUeST.

Figure 5-A.1. The vector bosons. Each circle represents a group generator. The known vector bosons are in the lowest row with a white interior. Yet to be found vector bosons are solid black. The Layer groups straddle all four layers. G1 is SU(3)⊗SU(2)⊗U(1). The list of groups for the higher three levels is the same as those of the first layer. There are 224 normal vector bosons not counting Species and Dark U(2) groups.

The "Dark" Vector Boson Spectrum

Layer 4

Layer 3

Layer 2

Dark Generation U(4)

Layer 1   **Dark** Group G2

Dark Layer U(4)
Dark Layer U(4)
Dark Layer U(4)
Dark Layer U(4)

Figure 5-A.2. The vector bosons. Each circle represents a group generator. The known vector bosons are in the lowest row with a white interior. Yet to be found vector bosons are solid black. The Layer groups are distributed by layer symbolically although they each straddle all four layers. G2 is $SU_D(2) \otimes U_D(1) \otimes SU_D(3)$. The list of groups for the higher three levels is the same as those of the first layer. There are 224 Dark vector bosons not counting Species and Dark U(2) groups.

5-A.4 Fermion Mass Spectrum

Figure 5-A.3. Fermion particle spectrum and partial example of pattern of mass mixing of the Generation group and of the Layer group. Unshaded parts are the known fermions with an additional, as yet not found, 4th generation. The lines on the left side (only shown for one layer) display the Generation mixing within each layer's species. The Generation mixing applies within each layer using a separate Generation group for each layer. The lines on the right side show Layer group mixing with the mixing amongst all four layers for each of the four generations individually. There are four Layer groups. There are 256 fundamental fermions. QUeST has the same fermion spectrum.

5-A.5 Groups and Fermion Splittings

Figure 5-A.4. The set of four layers of internal symmetry groups corresponding to four generations in four layers of spin ½ fermions and the four layers of vector bosons. In addition there are the Normal and Dark Layer groups, the Species group, and the Dark U(2) groups, which are *not* displayed.

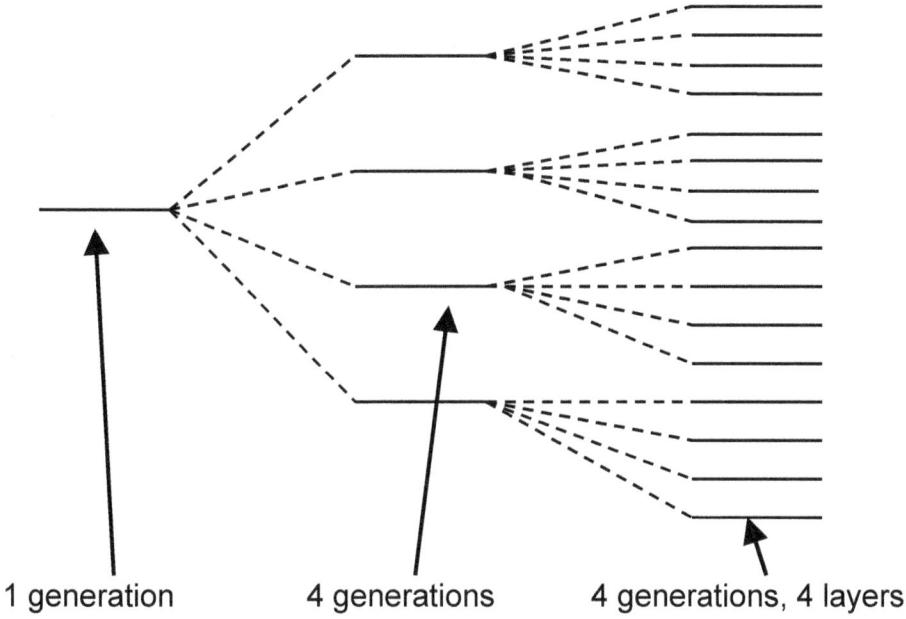

1 generation   4 generations   4 generations, 4 layers

Figure 5-A.5 The "splitting" of a single generation fermion into four generations and then into four layers.

Appendix 5-B. Four QUeST Fermion Species

QUeST is defined in a 3 + 1 dimensional complex quaternion space. These coordinates support a Complex Quaternion Lorentz Group just as the United SuperStandard theory. There are four types of boosts in the 3 + 1 Complex Quaternion space-time. Correspondingly, there are four species of QUeST fundamental fermions: "charged" lepton species, neutral lepton species, up-type quark species, and down-type quark species. The nature of each species is the same as in the Unified SuperStandard Theory.

The boosts are

1. Boosts from a rest frame to a rest frame with a relative velocity less than c.
2. Boosts from a rest frame to a rest frame with a relative velocity greater than c.
3. Boosts from a rest frame to a rest frame with a relative complex velocity less than c.
4. Boosts from a rest frame to a rest frame with a relative complex velocity greater than c.

The key relation for boosts in complex quaternion space is

$$e^x = e^a (\cos (\|\mathbf{v}\|) + \mathbf{v}/\|\mathbf{v}\| \sin(\|\mathbf{v}\|))s \qquad (5\text{-B}.1)$$

using definitions in Appendix 2-A, in analogy with the similar complex-valued identity used in Blaha (2020c).

6. Bioctonion Megaverse

Earlier we developed a biquaternion theory called QUeST that could be used to derive the internal symmetry group structure, and the fundamental fermion and vector boson spectrums. It led to the Unified SuperStandard Theory (UST) and accounted for the close similarity between the internal symmetries of the Standard Model sector and the subgroups of the Lorentz group. (They both exhibit U(1), SU(2) and SU(3) symmetries.)

6.1 MOST

If we assume the existence of a Megaverse[26] containing our universe, and other universes, then we can define a *Megaverse Octonion SuperStandard Theory* (MOST) that becomes a more general basis for QUeST and the Unified SuperStandard Theory.

In this chapter we define a bioctonion space and use it to define a Megaverse basis for MOST. MOST develops a more robust set of internal symmetries and fundamental particles. It creates a new view of Dark matter that appears to help explicate the lack of interactions between normal matter and Dark matter.

Remarkably MOST, when "restricted" to our universe, yields QUeST.

6.2 Motivation and Procedure

Our goal again is to create a larger dimension space within which a universe based on QUeST can exist, and where we can ultimately derive our space-time and the Unified SuperStandard Theory. Again we use the similarity of Lorentz subgroups and Standard Model internal symmetry groups in our development.

Again there are two possible procedures to follow in developing the deeper basis:

---

[26] The Megaverse is described in some detail in Blaha (2017c), (2017f), and (2018e) together with evidence for its existence.

1. One can develop the Quantum Mechanics, Quantum Field Theory, … in an octonion space and then extract the dynamics, fermion spectrum, gauge fields, and so on of our familiar space-time.

2. One can define an octonion space and then use its coordinates to directly extract the space-time, internal symmetries, fermion spectrum, gauge field spectrum and dynamics.

Again we have chosen the latter approach.

In developing the deeper space, upon which we build, we will take guidance from the derivation of the Unified SuperStandard Theory. This theory assumes a complex 4-dimensional space-time upon which Complex General Relativity is constructed. It then proceeds to complex flat space-time and Complex Relativity. Similarly, QUeST is based on a complex coordinate system: complex quaternion space.

After defining features of Complex Lorentz transformations the Unified SuperStandard Theory used Lorentz boosts to derive the Dirac forms of the four fermion species. The boosts were required to boost a fermion from a rest state to a state with a real-valued energy, and real-valued or complex-valued 3-momenta. Thus a real time –complex-valued spatial part is required for the proper definition of species.

Similarly we define a complex octonion (bioctonion) space, and use the real octonion time, and the complex octonion spatial part to define the subspace for internal symmetries.

As in the previous quaternion case (section 4.3), we will show that the MOST internal symmetry subgroups map to Standard Model internal symmetry subgroups.

6.3 Definition of Bioctonion Space

Following the above stated procedure we define a bioctonion[27] (complex octonion) space with *one* "time" biquaternion and *seven* "spatial" bioctonions as a generalization of the 3 + 1 space-time of our experience. The choice of 8 bioctonion dimensions seemed natural but was not required by a principle. It does lead to a larger

---

[27] We use bioctonion synonymously with complex octonion in this and subsequent chapters.

set of internal symmetries with *one layer* QUeST symmetries as a subset.[28] We will use the psiphi symbol • to represent each of the bioctonion space coordinates in Fig. 6.1.

We have chosen a complex 8-dimensional bioctonion space-time as the one MOST layer Megaverse space-time. There are 128 real coordinates in the bioctonion Megaverse space from which complex 8-dimensional real space-time (4-dimensional complex space-time) is extracted. Our universe's 4-dimensional complex space-time is a subspace-time.

Fig 6.1 symbolically depicts the space with a psiphi • for each real-valued coordinate. *Again we treat the bioctonion space as a higher dimensional space and do not use details of octonion algebra in our development.*

Time

• • • • • • • • • • • • • • • •

7-Space

• • • • • • • • • • • • • • • •
• • • • • • • • • • • • • • • •
• • • • • • • • • • • • • • • •
• • • • • • • • • • • • • • • •
• • • • • • • • • • • • • • • •
• • • • • • • • • • • • • • • •
• • • • • • • • • • • • • • • •

Figure 6.1. Eight-Dimensional (7 + 1) bioctonion space with coordinates represented by • 's. This bioctonion space has 128 real dimensions (64 complex dimensions.).

---

[28] We start the complex octonion discussion by developing a one layer MOST. Then we develop a four layer MOST which contains four layer QUeST-based universes.

6.4 Bioctonion Lorentz Group

Our definition of time and space bioctonion coordinates purposefully resembles those of our real space-time. One might ask why there should be a Lorentz-like group for bioctonion space.

The only apparent reason is the need for a special speed c whose existence enables one to boost from the rest frame of a mass m particle with rest energy m to a moving frame of greater energy.[29] Without c the group of the above coordinates would presumably be the bioctonion U(8) group. U(8) transformations preserve the norm of a state so that a "boost-like" transformation does not exist.

The bioctonion Lorentz transformations do have a unique speed c (the speed of light) and specify a unique rest frame for any particle—both sublight particles and tachyon particles.[30] . Thus we select the bioctonion Lorentz group for bioctonion space.

Flat space bioctonion Special Relativity generalizes to a bioctonion General Relativity which may be constructed directly (mindful of octonion non-commutativity).

Flat space bioctonion Lorentz group transformations have constant bioctonion matrix elements that are analogous to those of the Lorenz group.

---

[29] A similar consideration applies to tachyons.

[30] Tachyons can be transformed by a Complex Lorentz transformation to and from a rest frame. See Blaha (2018f).

7. Symmetries of the Megaverse Octonion SuperStandard Theory (MOST)

The coordinate space-time picture of the Megaverse described in chapter 6 enables us to simply find the internal symmetries and particle spectra of MOST. They will turn out to be a superset of those of the Unified SuperStandard Theory.

7.1 Extracting the Symmetries and Particle Spectra

As stated earlier in chapter 4 for the quaternion case, we will directly describe the symmetry structure implied by the form of the bioctonion (complex octonion) coordinate system.

The Unified SuperStandard Theory developed the group structure, from which the particle species were derived, from a subset of Lorentz boost transformations. Complex boosts mapped a system at rest to a system in motion with a real energy and complex 3-momenta in general. Bioctonion Lorentz boosts play a similar role.

The 4-dimensional representation of the Unified SuperStandard Theory complex coordinates subset is given in Fig. 7.1.

<u>**Time**</u>
•
<u>**Space**</u>
• •
• •
• •

Figure 7.1. Four-Dimensional space for Unified SuperStandard derivation of particle spectra with coordinates represented by psiphi • 's.

Following the same line of thought, which is described in chapter 4, we now specify a bioctonion subspace restricted to that of Fig. 7.2 to define the relevant set of coordinates for determining particle symmetries and spectra.

Time

• • • • • • • •

7-Space

• • • • • • • •   • • • • • • • •
• • • • • • • •   • • • • • • • •
• • • • • • • •   • • • • • • • •
• • • • • • • •   • • • • • • • •
• • • • • • • •   • • • • • • • •
• • • • • • • •   • • • • • • • •
• • • • • • • •   • • • • • • • •

Figure 7.2. 8 complex dimension bioctonion subspace for symmetries and particle spectra.

The subspace's 60 complex dimensions (coordinates) will split into an 8-dimensional complex coordinate space which supports 8-dimensional Complex Lorentz transformations, and a 52 complex dimension (coordinates) internal symmetry space. The neglected eight dimensions support a Dark U(4) rotations group[31] transforming between normal and Dark sectors. (There are four sectors in MOST necessitating the U(4) rotations that transform among them.)

The mechanism for this symmetry breakdown may be due to vacuum energy effects in the bioctonion Megaverse.

The MOST Megaverse subspace for the determination of the Internal Symmetries is depicted in Fig. 7.3. It is based on the discussion above.

---

[31] This Dark group is the analogue of the complex quaternion Dark U(2) group.

Figure 7.3. Complex 7 + 1 space-time (left box) and Internal Symmetry determination subspace (right box). The 16 real space-time coordinates are separated from 104 real coordinates for internal symmetries.

The internal symmetry coordinates above number 104 real coordinates or 52 complex coordinates. These coordinates serve as the coordinates of the fundamental representations of each of the factors of

$$[SU(2) \otimes U(1) \otimes SU(3) \otimes SU(2) \otimes U(1) \otimes SU(3)]^2 \otimes U(4)^8 \qquad (7.1)$$

The factorized internal symmetry emerges from another breakdown(s) which corresponds to the subgroup structure of the Lorentz group. Eq. 7.1 evidently follows from the structure of the bioctonion Lorentz transformations.

The U(4) Generation and Layer groups are represented in Fig. 7.1. We depict the pattern of symmetry implied by Fig. 7.1 and eq. 7.1 in Fig. 7.4 and 7.5 below.

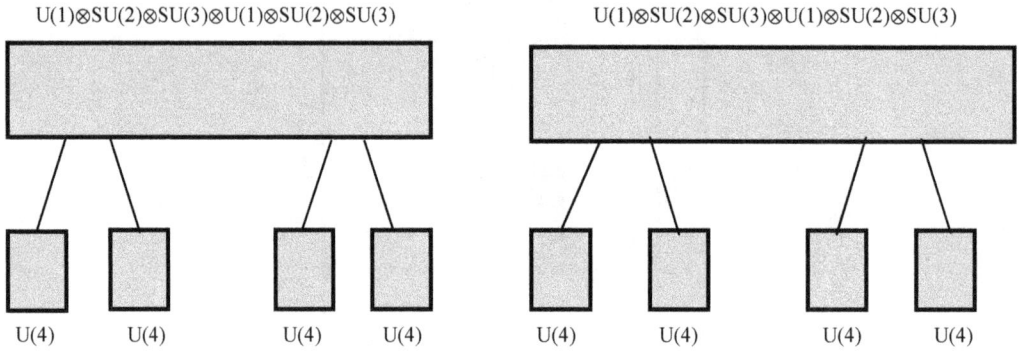

Figure 7.4. Schematic of the internal symmetry groups' coordinates of Fig. 7.1. The two "large" blocks are each sets of 20 real-valued coordinates furnishing representations of the indicated groups. The lower U(4) groups are the Generation and Layer number groups. The Dark U(4) group is not shown. The total number of real-valued coordinates is 104.

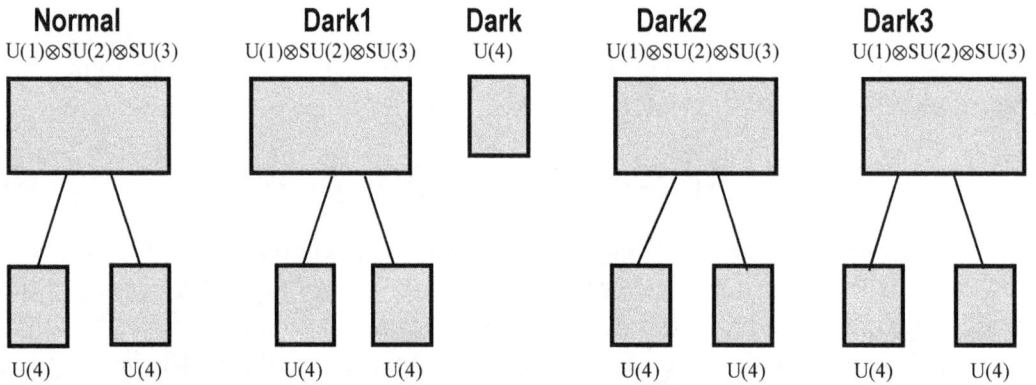

Figure 7.5. Schematic of the internal symmetry groups of eq. 7.1 with the Dark U(4) group added. These are the internal symmetry groups of one layer MOST. The lower U(4) groups are the Generation and Layer number groups. One pair of each number group is for each of the four $U(1) \otimes SU(2) \otimes SU(3)$ factors above.

The result is the total Internal Symmetry group of one layer of an generalization of QUeST and the Unified SuperStandard Theory.

Each $U(1)\otimes SU(2)\otimes SU(3)\otimes U(1)\otimes SU(2)\otimes SU(3)$ block in Fig. 7.4 has a 10 complex coordinates (20 real-valued coordinates) representation. The blocks are subdivided in Fig. 7.5 into sets of 10 real-valued coordinates supporting representations of $U(1)\otimes SU(2)\otimes SU(3)$. We assign the first block to contain the representations of the known parts of the Standard Model. There are three Dark blocks. The internal symmetry groups of each part are listed in Fig. 7.5.

The factorization of each of the four blocks is accomplished by following the procedure given in section 4.3 for each block.[32]

7.2 One Layer MOST

The above section specifies the *one layer* MOST. The symmetries of the three other layers are the same but their groups are individual to each layer. The groups of each layer are each flagged with a different index.

The overall one layer MOST internal symmetry is specified by Fig. 7.5.

$$[SU(2)\otimes U(1)\otimes SU(3)]^4\otimes U(4)^9 \tag{7.2}$$

There is also a space-time sector with 8 (= 7 + 1) complex coordinates.

Comparing Figs.4.2 and 7.5 we see that a one layer QUeST has a "Normal" part and one Dark part while the MOST adds two more Dark parts.

Thus the internal symmetry groups of one Layer MOST are:

"Normal" Gauge Groups
SU(3)⊗SU(2)⊗U(1)
Generation Group U(4)
Layer Group U(4)

---

[32] The separation of the 20 coordinate subspace into the subgroup factors representation can be implemented as group transformations and definitions using standard group theoretic methods. A more formal method for extracting the subgroup content of representations uses a symmetric group analysis of U(n) representation characters. See S. Blaha, J. Math. Phys. **10**, 2156 (1969) for a detailed discussion of this approach.

Dark1 Gauge Groups
SU(3)⊗SU(2)⊗U(1)
Generation Group U(4)
Layer Group U(4)

Dark2 Gauge Groups
SU(3)⊗SU(2)⊗U(1)
Generation Group U(4)
Layer Group U(4)

Dark3 Gauge Groups
SU(3)⊗SU(2)⊗U(1)
Generation Group U(4)
Layer Group U(4)

PLUS

A Dark U(4) group that rotates among the four normal and Dark sectors

Figure 7.6. One layer MOST vector bosons list from eq. 7.2. The four layer MOST quadruples the above list: with one distinct set for each layer. In one layer the total number of vector bosons of the above list is 192 for one layer. Thus four layers yield a total count of 768 vector bosons in MOST (not counting the Species group which comes from General Relativity). We require each level has a separate Dark U(4) rotation group.

7.3 Four Layer MOST

The four layer MOST is described by a 32 dimension complex octonion space. Thus it consists of four "copies" of the coordinates:

Figure 7.7. The 32 dimension MOST schematic. Four layer MOST has 512 real-valued dimensions.

which yield four duplicates of the internal symmetry schematic in Fig. 7.5,

and an 8 complex quaternion space-time consisting of 7 + 1 complex-valued quaternion coordinates.

The sum total of real-valued dimensions is 512 as is the sum of the dimensions of the above parts constructed from the dimensions.

> *In short we obtain a four layer MOST*
> *which may contain QUeST universes.*

7.4 Justification for a Four Layer MOST

There is good reason for MOST to have four layers embodied in 32 dimension complex quaternion space. If one considers the content of a layer displayed in Fig. 7.3 one sees a 8-dimension complex coordinates block for space-time. To create an 8 dimension complex *quaternion* coordinates space-time, one needs four layers of the form of Fig. 7.3. *The combination of four 8-dimension complex coordinates is an 8 dimension complex quaternion space-time.*

Thus the choice of four layer MOST gives us an 8-dimension complex quaternion space-time that can contain 4-dimension complex quaternion QUeST universes. *We conclude four layer MOST is needed to have an 8-dimension complex quaternion space-time.*

The thirty-two dimension complex octonion space contains an 8-dimension complex quaternion space-time and the four layers of Internal Symmetry groups shown in Fig. 7.5.

7.5 Fermion and Gauge Vector Boson Spectrums

The fermion and vector boson spectrums that emerge in MOST are those of an "enlarged" QUeST and Unified SuperStandard Theory. They are displayed below. MOST has an additional two Dark sectors beyond QUeST and the Unified SuperStandard Theory.

Vector Bosons

From Fig. 7.6 we find MOST has 192 vector bosons in one layer. Thus four layer MOST has a total count of 768 MOST vector bosons. There are two additional Dark vector boson sectors beyond QUeST and the Unified SuperStandard Theory.

Fermions

There are 512 fundamental fermions in MOST, which includes two additional Dark fermion sectors. Fig. 7.8 shows the MOST fermion spectrum.

```
        1      2      3      4
     ....  ....  ....  ....
     ....  ....  ....  ....
     ....  ....  ....  ....
     ....  ....  ....  ....

     ....  ....  ....  ....
     ....  ....  ....  ....
     ....  ....  ....  ....
     ....  ....  ....  ....

     ....  ....  ....  ....
     ....  ....  ....  ....
     ....  ....  ....  ....
     ....  ....  ....  ....

     ....  ....  ....  ....
     oooo  ....  ....  ....
     oooo  ....  ....  ....
     oooo  ....  ....  ....
```

Figure 7.8. Schematic spectrum of the fermions of 4 layer MOST. Each fermion is represented by a •. Quark triplets are represented by a single •. Four sets of four species in four generations which are in turn in 4 layers. Open symbols ◦ represent known fermions. There are 512 fundamental fermions taking account of quark triplets. Note the Layer groups determine the layers in UST. **They require 4 layers of 8 complex octonions in Megaverse space leading to the 32 dimension complex octonion space.**

8. The MOST Megaverse

The MOST Megaverse *space-time* has 8 complex quaternion dimensions (64 real-valued dimensions counting the 8 dimensions within each of the eight quaternion coordinates). It has a larger set of Internal Symmetries and particles than a QUeST universe. Four layer QUeST universes can exist in a four layer MOST Megaverse.

MOST has four $U(1) \otimes SU(2) \otimes SU(3)$ parts (Fig. 7.5) with a different set of $U(1) \otimes SU(2) \otimes SU(3)$ gauge vector bosons in each part. It also has four blocks of fermions (shown in Fig. 7.8).

8.1 QUeST Universes Within MOST

Given the four parts of MOST Internal Symmetries the question arises as to how a QUeST universe, which has only two parts, is constructed. The four parts have the same groups, and the same associated fermion and vector boson particles. The author has shown that the coupling constants of the groups are determined by vacuum polarization effects. (See Blaha (2020c).) The remaining features: particle masses and symmetry breaking, are determined by vacuum effects, which suggests that they may be the same for all four parts of MOST.[33]

If we construct a universe (conceptually) there are 6 possible combinations of the four parts, from which two parts are selected. If all parts are the same, as the previous paragraph suggests, then all combinations lead to the same QUeST universe. If the parts differ then there would be up to 6 variants of QUeST universes. Our universe then becomes one possibility.

---

[33] In the Megaverse, the Megaverse vacuum determines these quantities. In a universe, its vacuum determines these quantities.

```
┌─────────────────────┐
│                     │
│        MOST         │      7 + 1 Complex Quaternion
│                     │
└─────────────────────┘
           │
           ▼
┌─────────────────────┐
│                     │
│        QUeST        │      3 + 1 Complex Quaternion
│                     │
└─────────────────────┘
           │
           ▼
┌─────────────────────┐
│      Unified        │
│   SuperStandard     │      3 + 1 Real-Valued
│      Theory         │
└─────────────────────┘
```

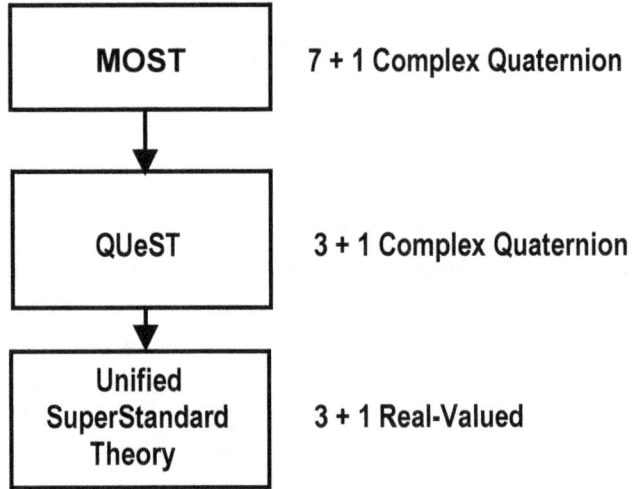

Figure 8.1. The hierarchy of Unified SuperStandard Theories.

8.2 MOST Spinors

Note the 7 + 1 Complex Octonion Megaverse's space-time has 16 component Dirac fermion spinors.

8.3 MOST Fermion Species

The fermion species in MOST number four despite the complex eight-dimensional nature of the space-time extracted from complex octonion space. The eight complex space-time coordinates support a 7+1 Complex Quaternion Lorentz Group.

There are four types of boosts in this 7+1 complex quaternion space-time. They correspond to the types of boosts in UST. They boost a particle rest state to a state of motion with a real energy p^0 (the magnitude of the quaternion-valued energy), and a real quaternion valued or complex quaternion valued momentum (spatial coordinates) whose magnitude $|\mathbf{p}|$ is a real-valued or complex-valued quantity. The fermions that can be generated by $7 + 1$ Complex Quaternion Lorentz Group boosts[34] are:

1. A boost from rest to a frame with real-valued energy and momentum with $p^{02} - \mathbf{p}^2 > 0$. A "normal charged lepton-like" fermion.

2. A boost from rest to a frame with real-valued energy and momentum with $p^{02} - \mathbf{p}^2 < 0$. A "tachyonic neutral lepton-like" fermion.

3. A boost from rest to a frame with real-valued energy and a complex-valued spatial momentum with $p^{02} - \mathbf{p}^2 > 0$. An "up-type quark-like" fermion.

4. A boost from rest to a frame with real-valued energy and momentum with $p^{02} - \mathbf{p}^2 < 0$. A "tachyonic down-type quark-like" fermion.

where p^0 is the energy magnitude and $\mathbf{p}$ is a spatial 7-vector magnitude. The particle states generated by 1 and 2 are "conventional" 8-dimensional analogues of the 4-dimensional case.

8.4 Octoquarks

The quark-like cases have an 8-dimensional aspect that prompts us to call them *octoquarks*. In 4-dimensional complex space-time a quark momentum is set by the term

$$\omega w = \mathbf{u}_r \omega_r + i\mathbf{u}_i \omega_i \qquad (8.1)$$

as shown in Blaha (2020c). We now limit the 8-dimensional complex quaternion space-time to a complex space-time. In 8-dimensional complex space-time the 8-momentum, the *hyper-momentum*,[35] is set by

---

[34] The boosts rely on eq. 2-A.4.
[35] As opposed to complex momentum in 4-space-time.

$$\omega w = u_r \omega_r + i u_r \omega_1 + j u_2 \omega_2 + k u_3 \omega_3 + q u_4 \omega_4 + r u_5 \omega_5 + s u_6 \omega_6 \qquad (8.2)$$

where the u_i are 8-vectors satisfying

$$u_i \cdot u_j = \delta_{ij} \qquad (8.3)$$

and i, j, k, q, r, and s are fundamental quaternion-like units.

Note that a fermion spin s satisfies the 8-vector inner product

$$s \cdot p = 0 \qquad (8.4)$$

or

$$s \cdot p = 0 \qquad (8.5)$$

in a manner similar to 4-space-time.

Thus the seven 7-vectors in 8-space-time form an orthonormal set and define quarks with 7-momentum. We call them *octoquarks* since the space-time is 8-dimensional.

Normal and tachyonic particles are distinguished in complex 8-dimensional space by the sign of $p^{02} - p^2$ in a manner similar to 4-space-time. MOST fermions occur in four species just like QUeST fermions and Unified SuperStandard Theory fermions.

There are four species of fundamental fermions in 4-space-time and 8-space-time: "charged" lepton species, neutral lepton species, up-type octoquark species, and down-type octoquark species. The nature of each species (normal or tachyon, and real-valued or complex spatial momentum) is the same as in the Unified SuperStandard Theory.

9. Particle-Dimension Duality

In 32 dimension complex quaternion space there are 256 real dimensions and 256 fundamenatl fermions. In 32 dimension complex octonion space there are 512 real dimensions and 512 fundamenatl fermions. The equality of these parameters raises the possibility that the number of space dimensions equals the number of fundamental fermions in some deep sense. This possibility is enhanced because the number of particles depends on the singlets nature of leptons and the triplet nature of quarks.

When one considers the measurement of distance in a space one sees the measuring process requires the use of particles. Without particles one cannot distinguish space-time points from each other. And the distance between space-time points is determined either directly or indirectly by bosons such as light transmitted (and possibly reflected) between fermion "clumps.'

Without particles distance, space-time is meaningless. Without space-time particles will "clump' and dynamics is impossible.

Thus a relation between particles and dimensions is not unnatural in Physics.

9.1 Particles Mapped to Quantum Functionals

A particle has a quantum field. We have shown in earlier books that quantum fields may be factored into an inner product of a quantum functional and a wave in coordinate space-time.[36] Every particle has a set of internal symmetry and space-time quantum numbers. We can map the set of fundamental fermions to a set of quantum functionals. These functionals have transformation properties under internal symmetries. They are thus representations of the internal symmetry groups.

There is no distance in the set of functionals. As we showed in Blaha (2020) and earlier books factoring quantum fields into inner products of functionals and waves eliminates the instantaneity problem of Quantum Entanglement.

---

[36] See Blaha (2019e) and (2020) as well as earlier books by the author.

9.2 Coordinates and Functionals

The lack of distance in the set of functionals parallels the lack of distance between coordinates. For example there is no distance between the "x" coordinate and the "y" coordinate in a coordinate system. Functionals have features in common with coordinates.

Quantum functionals can also have commutation relations similar to coordinates: Defining a quantum functional conjugate momentum:

$$\pi_k = \, d/df_k \tag{9.1}$$

For quantum functional f_k we obtain the commutation relation

$$[f_k, \pi_j] = i\delta_{kj} \tag{9.2}$$

where k and j represent internal symmetry indices. Eq. 9.2 mirrors the form of quantum mechanical commutation relations adding to the similarity between coordinates (dimensions) and quantum functionals.

Biquaternion and bioctonion coordinates can be mapped to quantum functionals since they have the same role, and number, as the respective coordinates. We may set

$$x_m = R_{mn}f_n \tag{9.3}$$

where R is a transformation.

9.3 Particle-Dimension Duality

Thus in QUeST and MOST we find a particle-dimension (coordinate) duality. The point of view offered by this duality suggests that fermion particles are in some sense "interchangeable" with coordinates—with particle functionals as an intermediary.

9.4 Particle – Functional – Dimension Triality

The above discussion shows a more general analogy – a *triality* – between fermion particles, dimensions (coordinates) and quantum functionals.

9.5 Functionals Implement Quantum Entanglement

In 1935 Einstein, Podolsky, and Rosen[37] (EPR) raised questions about Quantum Entanglement.

9.5.1 Quantum Entanglement and Action-at-a-Distance

EPR considered the quantum entanglement of two systems and showed that instantaneous action-at-a-distance (spookiness) resulted. In this section we will show that our quantum functional formalism, which generalizes quantum theory, eliminates the problem of instantaneous action-at-a-distance.[38] We will show the solution provided by quantum functionals using the same example as EPR.

The key feature of quantum functionals is their ubiquitous presence at every point of space-time. In a multi-system quantum state all functionals are directly, instantaneously linked no matter what the separation of the constituent systems. When a reduction of the state of one system occurs due to a measurement, all other systems are instantly updated since the space-time separation of the individual systems is not relevant. The linkage of all quantum functionals is relevant. A reduction of one system immediately impacts the other related systems.

9.6 The EPR Two System State Example

EPR considered a state consisting of two systems that might become separated spatially. We can represent the state as

$$\Psi = \Sigma_n \psi_{1n}(x_1)\psi_{2n}(x_2) \tag{9.4}$$

We can represent a measurement (reduction of state) with a projection Π_{1a} of system "1" to a state ψ_{1a} with

$$\psi_{1a} = \delta_{ab} \Pi_a \psi_{1b} \tag{9.5}$$

---

[37] Einstein A, Podolsky B, and Rosen N, "Can Quantum-Mechanical Description of Physical Reality Be Considered Complete?", Phys. Rev. **47**, 777 (1935).
[38] This section presents the solution for quantum spookiness that we proposed in Blaha (2019g) and (2018e).

Then

$$\Psi_{\text{projected}} = \Pi_{1a} \Sigma_n \psi_{1n} \psi_{2n} = \psi_{1a}(x_1) \psi_{2a}(x_2) \tag{9.6}$$

The effect of the measurement of system "1" is *instantaneous* of system "2" because the quantum functionals f_{1n} and f_{2n}, and the projections Π_{1n} and Π_{2n} of both systems are not separated by distance with the result

$$\psi_{1n}(x) = f_{1xn}(\Pi_{1xn}\Phi) = (f_{1xn}, \Pi_{1xn}\Phi) \tag{9.7}$$

$$\psi_{2n}(y) = f_{2ny}(\Pi_{2yn}\Phi) = (f_{2yn}, \Pi_{2yn}\Phi) \tag{9.8}$$

with $x = x_1$ and $y = x_2$. *The quantum functional and the projection select the wave and its coordinate parameterization. The coordinates in the wave are merely place holders.*

Therefore the relative distance between the coordinates x_1 and x_2 is not relevant for the change of state of system "2". The quantum functionals and projections give the instantaneity of the change in ψ_{2a} upon the measurement of system "1".

The EPR Spookiness is resolved by quantum functionals. There is no conflict with the Theory of Special Relativity.

REFERENCES

Akhiezer, N. I., Frink, A. H. (tr), 1962, *The Calculus of Variations* (Blaisdell Publishing, New York, 1962).

Bjorken, J. D., Drell, S. D., 1964, *Relativistic Quantum Mechanics* (McGraw-Hill, New York, 1965).

Bjorken, J. D., Drell, S. D., 1965, *Relativistic Quantum Fields* (McGraw-Hill, New York, 1965).

Blaha, S., 1998, *Cosmos and Consciousness* (Pingree-Hill Publishing, Auburn, NH, 1998).

_____, 2002, *A Finite Unified Quantum Field Theory of the Elementary Particle Standard Model and Quantum Gravity Based on New Quantum Dimensions™ & a New Paradigm in the Calculus of Variations* (Pingree-Hill Publishing, Auburn, NH, 2002).

_____, 2003, *A Finite Unified Quantum Field Theory of the Elementary Particle Standard Model and Quantum Gravity Based on New Quantum Dimensions™ and a New Paradigm in the Calculus of Variations* (Pingree-Hill Publishing, Auburn, NH, 2003).

_____, 2004, *Quantum Big Bang Cosmology: Complex Space-time General Relativity, Quantum Coordinates™Dodecahedral Universe, Inflation, and New Spin 0, ½, 1 & 2 Tachyons & Imagyons* (Pingree-Hill Publishing, Auburn, NH, 2004).

_____, 2005a, *Quantum Theory of the Third Kind: A New Type of Divergence-free Quantum Field Theory Supporting a Unified Standard Model of Elementary Particles and Quantum Gravity based on a New Method in the Calculus of Variations* (Pingree-Hill Publishing, Auburn, NH, 2005).

_____, 2005b, *The Metatheory of Physics Theories, and the Theory of Everything as a Quantum Computer Language* (Pingree-Hill Publishing, Auburn, NH, 2005).

_____, 2005c, *The Equivalence of Elementary Particle Theories and Computer Languages: Quantum Computers, Turing Machines, Standard Model, Superstring Theory, and a Proof that Gödel's Theorem Implies Nature Must Be Quantum* (Pingree-Hill Publishing, Auburn, NH, 2005).

_____, 2006a, *The Foundation of the Forces of Nature* (Pingree-Hill Publishing, Auburn, NH, 2006).

_____, 2006b, *A Derivation of ElectroWeak Theory based on an Extension of Special Relativity; Black Hole Tachyons; & Tachyons of Any Spin.* (Pingree-Hill Publishing, Auburn, NH, 2006).

_____, 2007a, *Physics Beyond the Light Barrier: The Source of Parity Violation, Tachyons, and A Derivation of Standard Model Features* (Pingree-Hill Publishing, Auburn, NH, 2007).

_____, 2007b, *The Origin of the Standard Model: The Genesis of Four Quark and Lepton Species, Parity Violation, the ElectroWeak Sector, Color SU(3), Three Visible Generations of Fermions, and One Generation of Dark Matter with Dark Energy* (Pingree-Hill Publishing, Auburn, NH, 2007).

_____, 2008a, *A Direct Derivation of the Form of the Standard Model From GL(16) (Pingree-Hill Publishing, Auburn, NH, 2008).*

_____, 2008b, *A Complete Derivation of the Form of the Standard Model With a New Method to Generate Particle Masses Second Edition* (Pingree-Hill Publishing, Auburn, NH, 2008)

_____, 2009, *The Algebra of Thought & Reality: The Mathematical Basis for Plato's Theory of Ideas, and Reality Extended to Include A Priori Observers and Space-Time Second Edition* (Pingree-Hill Publishing, Auburn, NH, 2009).

_____, 2010a, *Operator Metaphysics: A New Metaphysics Based on a New Operator Logic and a New Quantum Operator Logic that Lead to a Mathematical Basis for Plato's Theory of Ideas and Reality* (Pingree-Hill Publishing, Auburn, NH, 2010).

_____, 2010b, *The Standard Model's Form Derived from Operator Logic, Superluminal Transformations and GL(16)* (Pingree-Hill Publishing, Auburn, NH, 2010).

_____, 2010c, *SuperCivilizations: Civilizations as Superorganisms* (McMann-Fisher Publishing, Auburn, NH, 2010).

_____, 2011a, *21st Century Natural Philosophy Of Ultimate Physical Reality* (McMann-Fisher Publishing, Auburn, NH, 2011).

_____, 2011b, *All the Universe! Faster Than Light Tachyon Quark Starships & Particle Accelerators with the LHC as a Prototype Starship Drive Scientific Edition* (Pingree-Hill Publishing, Auburn, NH, 2011).

_____, 2011c, *From Asynchronous Logic to The Standard Model to Superflight to the Stars* (Blaha Research, Auburn, NH, 2011).

_____, 2012a, *From Asynchronous Logic to The Standard Model to Superflight to the Stars volume 2: Superluminal CP and CPT, U(4) Complex General Relativity and The Standard Model, Complex Vierbein General Relativity, Kinetic Theory, Thermodynamics* (Blaha Research, Auburn, NH, 2012).

_____, 2012b, *Standard Model Symmetries, And Four And Sixteen Dimension Complex Relativity; The Origin Of Higgs Mass Terms* (Blaha Reasearch, Auburn, NH, 2012).

_____, 2013a, *Multi-Stage Space Guns, Micro-Pulse Nuclear Rockets, and Faster-Than-Light Quark-Gluon Ion Drive Starships* (Blaha Research, Auburn, NH, 2013).

_____, 2013b, *The Bridge to Dark Matter; A New Sister Universe; Dark Energy; Inflatons; Quantum Big Bang; Superluminal Physics; An Extended Standard Model Based on Geometry* (Blaha Reasearch, Auburn, NH, 2013).

_____, 2014a, *Universes and Megaverses: From a New Standard Model to a Physical Megaverse; The Big Bang; Our Sister Universe's Wormhole; Origin of the Cosmological Constant, Spatial Asymmetry of the Universe, and its Web of Galaxies; A Baryonic Field between Universes and Particles; Megaverse Extended Wheeler-DeWitt Equation* (Blaha Reasearch, Auburn, NH, 2014).

_____, 2014b, *All the Megaverse! Starships Exploring the Endless Universes of the Cosmos Using the Baryonic Force* (Blaha Research, Auburn, NH, 2014).

_____, 2014c, *All the Megaverse! II Between Megaverse Universes: Quantum Entanglement Explained by the Megaverse Coherent Baryonic Radiation Devices – PHASERs Neutron Star Megaverse Slingshot Dynamics Spiritual and UFO Events, and the Megaverse Microscopic Entry into the Megaverse* (Blaha Research, Auburn, NH, 2014).

_____, 2015a, *PHYSICS IS LOGIC PAINTED ON THE VOID: Origin of Bare Masses and The Standard Model in Logic, U(4) Origin of the Generations, Normal and Dark Baryonic Forces, Dark Matter, Dark Energy, The Big Bang, Complex General Relativity, A Megaverse of Universe Particles* (Blaha Research, Auburn, NH, 2015).

_____, 2015b, *PHYSICS IS LOGIC Part II: The Theory of Everything, The Megaverse Theory of Everything, U(4)⊗U(4) Grand Unified Theory (GUT), Inertial Mass = Gravitational Mass, Unified Extended Standard Model and a New Complex General Relativity with Higgs Particles, Generation Group Higgs Particles* (Blaha Research, Auburn, NH, 2015).

_____, 2015c, *The Origin of Higgs ("God") Particles and the Higgs Mechanism: Physics is Logic III, Beyond Higgs – A Revamped Theory With a Local Arrow of Time, The Theory of Everything Enhanced, Why Inertial Frames are Special, Universes of the Mind* (Blaha Research, Auburn, NH, 2015).

_____, 2015d, *The Origin of the Eight Coupling Constants of The Theory of Everything: U(8) Grand Unified Theory of Everything (GUTE), S^8 Coupling Constant Symmetry, Space-Time Dependent Coupling Constants, Big Bang Vacuum Coupling Constants, Physics is Logic IV* (Blaha Research, Auburn, NH, 2015).

_____, 2016a, *New Types of Dark Matter, Big Bang Equipartition, and A New U(4) Symmetry in the Theory of Everything: Equipartition Principle for Fermions, Matter is 83.33% Dark,*

Penetrating the Veil of the Big Bang, Explicit QFT Quark Confinement and Charmonium, Physics is Logic V (Blaha Research, Auburn, NH, 2016).

_____, 2016b, *The Periodic Table of the 192 Quarks and Leptons in The Theory of Everything: The U(4) Layer Group, Physics is Logic VI* (Blaha Research, Auburn, NH, 2016).

_____, 2016c, *New Boson Quantum Field Theory, Dark Matter Dynamics, Dark Matter Fermion Layer Mixing, Genesis of Higgs Particles, New Layer Higgs Masses, Higgs Coupling Constants, Non-Abelian Higgs Gauge Fields, Physics is Logic VII* (Blaha Research, Auburn, NH, 2016).

_____, 2016d, *Unification of the Strong Interactions and Gravitation: Quark Confinement Linked to Modified Short-Distance Gravity; Physics is Logic VIII* (Blaha Research, Auburn, NH, 2016).

_____, 2016e, *MoND: Unification of the Strong Interactions and Gravitation II, Quark Confinement Linked to Large-Scale Gravity, Physics is Logic IX* (Blaha Research, Auburn, NH, 2016).

_____, 2016f, *CQ Mechanics: A Unification of Quantum & Classical Mechanics, Quantum/Semi-Classical Entanglement, Quantum/Classical Path Integrals, Quantum/Classical Chaos* (Blaha Research, Auburn, NH, 2016).

_____, 2016g, *GEMS: Unified Gravity, ElectroMagnetic and Strong Interactions: Manifest Quark Confinement, A Solution for the Proton Spin Puzzle, Modified Gravity on the Galactic Scale* (Pingree Hill Publishing, Auburn, NH, 2016).

_____, 2016h, *Unification of the Seven Boson Interactions based on the Riemann-Christoffel Curvature Tensor* (Pingree Hill Publishing, Auburn, NH, 2016).

_____, 2017a, *Unification of the Eleven Boson Interactions based on 'Rotations of Interactions'* (Pingree Hill Publishing, Auburn, NH, 2017).

_____, 2017b, *The Origin of Fermions and Bosons, and Their Unification* (Pingree Hill Publishing, Auburn, NH, 2017).

_____, 2017c, *Megaverse: The Universe of Universes* (Pingree Hill Publishing, Auburn, NH, 2017).

_____, 2017d, *SuperSymmetry and the Unified SuperStandard Model* (Pingree Hill Publishing, Auburn, NH, 2017).

_____, 2017e, *From Qubits to the Unified SuperStandard Model with Embedded SuperStrings: A Derivation* (Pingree Hill Publishing, Auburn, NH, 2017).

_____, 2017f, *The Unified SuperStandard Model in Our Universe and the Megaverse: Quarks, ... ,* (Pingree Hill Publishing, Auburn, NH, 2017).

_____, 2018a, *The Unified SuperStandard Model and the Megaverse SECOND EDITION A Deeper Theory based on a New Particle Functional Space that Explicates Quantum Entanglement Spookiness (Volume 1)* (Pingree Hill Publishing, Auburn, NH, 2018).

_____, 2018b, *Cosmos Creation: The Unified SuperStandard Model, Volume 2, SECOND EDITION* (Pingree Hill Publishing, Auburn, NH, 2018).

_____, 2018c, *God Theory (*Pingree Hill Publishing, Auburn, NH, 2018).

_____, 2018d, *Immortal Eye: God Theory: Second Edition* (Pingree Hill Publishing, Auburn, NH, 2018).

_____, 2018e, *Unification of God Theory and Unified SuperStandard Model THIRD EDITION* (Pingree Hill Publishing, Auburn, NH, 2018).

_____, 2019a, *Calculation of: QED α = 1/137, and Other Coupling Constants of the Unified SuperStandard Theory* (Pingree Hill Publishing, Auburn, NH, 2019).

_____, 2019b, *Coupling Constants of the Unified SuperStandard Theory SECOND EDITION* (Pingree Hill Publishing, Auburn, NH, 2019).

_____, 2019c, *New Hybrid Quantum Big_Bang–Megaverse_Driven Universe with a Finite Big Bang and an Increasing Hubble Constant* (Pingree Hill Publishing, Auburn, NH, 2019).

_____, 2019d, *The Universe, The Electron and The Vacuum* (Pingree Hill Publishing, Auburn, NH, 2019).

_____, 2019e, *Quantum Big Bang – Quantum Vacuum Universes (Particles)* (Pingree Hill Publishing, Auburn, NH, 2019).

_____, 2019f, *The Exact QED Calculation of the Fine Structure Constant Implies ALL 4D Universes have the Same Physics/Life Prospects* (Pingree Hill Publishing, Auburn, NH, 2019).

_____, 2019g, *Unified SuperStandard Theory and the SuperUniverse Model: The Foundation of Science* (Pingree Hill Publishing, Auburn, NH, 2019).

_____, 2020a, *Quaternion Unified SuperStandard Theory (The QUeST) and Megaverse Octonion SuperStandard Theory (MOST)* (Pingree Hill Publishing, Auburn, NH, 2020).

_____, 2020b, *United Universes Quaternion Universe - Octonion Megaverse* (Pingree Hill Publishing, Auburn, NH, 2020).

_____, 2020c, *Unified SuperStandard Theories for Quaternion Universes & The Octonion Megaverse* (Pingree Hill Publishing, Auburn, NH, 2020).

Eddington, A. S., 1952, *The Mathematical Theory of Relativity* (Cambridge University Press, Cambridge, U.K., 1952).

Fant, Karl M., 2005, *Logically Determined Design: Clockless System Design With NULL Convention Logic* (John Wiley and Sons, Hoboken, NJ, 2005).

Feinberg, G. and Shapiro, R., 1980, *Life Beyond Earth: The Intelligent Earthlings Guide to Life in the Universe* (William Morrow and Company, New York, 1980).

Gelfand, I. M., Fomin, S. V., Silverman, R. A. (tr), 2000, *Calculus of Variations* (Dover Publications, Mineola, NY, 2000).

Giaquinta, M., Modica, G., Souchek, J., 1998, *Cartesian Coordinates in the Calculus of Variations* Volumes I and II (Springer-Verlag, New York, 1998).

Giaquinta, M., Hildebrandt, S., 1996, *Calculus of Variations* Volumes I and II (Springer-Verlag, New York, 1996).

Gradshteyn, I. S. and Ryzhik, I. M., 1965, *Table of Integrals, Series, and Products* (Academic Press, New York, 1965).

Heitler, W., 1954, *The Quantum Theory of Radiation* (Claendon Press, Oxford, UK, 1954).

Huang, Kerson, 1992, *Quarks, Leptons & Gauge Fields 2nd Edition* (World Scientific Publishing Company, Singapore, 1992).

Jost, J., Li-Jost, X., 1998, *Calculus of Variations* (Cambridge University Press, New York, 1998).

Kaku, Michio, 1993, *Quantum Field Theory*, (Oxford University Press, New York, 1993).

Kirk, G. S. and Raven, J. E., 1962, *The Presocratic Philosophers* (Cambridge University Press, New York, 1962).

Landau, L. D. and Lifshitz, E. M., 1987, *Fluid Mechanics 2nd Edition*, (Pergamon Press, Elmsford, NY, 1987).

Misner, C. W., Thorne, K. S., and Wheeler, J. A., 1973, *Gravitation* (W. H. Freeman, New York, 1973).

Rescher, N., 1967, *The Philosophy of Leibniz* (Prentice-Hall, Englewood Cliffs, NJ, 1967).

Rieffel, Eleanor and Polak, Wolfgang, 2014, *Quantum Computing* (MIT Press, Cambridge, MA, 2014).

Riesz, Frigyes and Sz.-Nagy, Béla, 1990, *Functional Analysis* (Dover Publications, New York, 1990).

Sagan, H., 1993, *Introduction to the Calculus of Variations* (Dover Publications, Mineola, NY, 1993).

Sakurai, J. J., 1964, *Invariance Principles and Elementary Particles* (Princeton University Press, Princeton, NJ, 1964).

Sorokin, Pitirim, 1941, *Social and Cultural Dynamics* (Porter Sargent Publishers, Boston, MA, 1941).

Streater, R. F. and Wightman, A. S., 2000, *PCT, Spin, Statistics, and All That* (Princeton University Press, Princeton, NJ 2000).

Weinberg, S., 1972, *Gravitation and Cosmology* (John Wiley and Sons, New York, 1972).

Weinberg, S., 1995, *The Quantum Theory of Fields Volume I* (Cambridge University Press, New York, 1995).

Weinberg, S., 2000, *The Quantum Theory of Fields Volume III Supersymmetry* (Cambridge University Press, New York, 2000).

Weyl, H., 1950, *Space, Time, Matter* (Dover, New York, 1950).

Weyl, H., (Tr. S. Pollard et al), 1987, *The Continuum* (Dover Publications, New York, 1987).

INDEX

About the Author

Stephen Blaha is a well-known Physicist and Man of Letters with interests in Science, Society and civilization, the Arts, and Technology. He had an Alfred P. Sloan Foundation scholarship in college. He received his Ph.D. in Physics from Rockefeller University. He has served on the faculties of several major universities. He was also a Member of the Technical Staff at Bell Laboratories, a manager at the Boston Globe Newspaper, a Director at Wang Laboratories, and President of Blaha Software Inc. and of Janus Associates Inc. (NH).

Among other achievements he was a co-discoverer of the "r potential" for heavy quark binding developing the first (and still the only demonstrable) non-Aeolian gauge theory with an "r" potential; first suggested the existence of topological structures in superfluid He-3; first proposed Yang-Mills theories would appear in condensed matter phenomena with non-scalar order parameters; first developed a grammar-based formalism for quantum computers and applied it to elementary particle theories; first developed a new form of quantum field theory without divergences (thus solving a major 60 year old

problem that enabled a unified theory of the Standard Model and Quantum Gravity without divergences to be developed); first developed a formulation of complex General Relativity based on analytic continuation from real space-time; first developed a generalized non-homogeneous Robertson-Walker metric that enabled a quantum theory of the Big Bang to be developed without singularities at t = 0; first generalized Cauchy's theorem and Gauss' theorem to complex, curved multi-dimensional spaces; received Honorable Mention in the Gravity Research Foundation Essay Competition in 1978; first developed a physically acceptable theory of faster-than-light particles; first derived a composition of extremums method in the Calculus of Variations; first quantitatively suggested that inflationary periods in the history of the universe were not needed; first proved Gödel's Theorem implies Nature must be quantum; provided a new alternative to the Higgs Mechanism, and Higgs particles, to generate masses; first showed how to resolve logical paradoxes including Gödel's Undecidability Theorem by developing Operator Logic and Quantum Operator Logic; first developed a quantitative harmonic oscillator-like model of the life cycle, and interactions, of civilizations; first showed how equations describing superorganisms also apply to civilizations. A recent book shows his theory applies successfully to the past 14 years of history and to *new* archaeological data on Andean and Mayan civilizations as well as Early Anatolian and Egyptian civilizations.

He first developed an axiomatic derivation of the form of The Standard Model from geometry – space-time properties – The Unified SuperStandard Model. It unifies all the known forces of Nature. It also has a Dark Matter sector that includes a Dark ElectroWeak sector with Dark doublets and Dark gauge interactions. It uses quantum coordinates to remove infinities that crop up in most interacting quantum field theories and additionally to remove the infinities that appear in the Big Bang and generate inflationary growth of the universe. It shows gravity has a MOND-like form without sacrificing Newton's Laws. It relates the interactions of the MOND-like sector of gravity with the r-potential of Quark Confinement. The axioms of the theory lead to the question of their origin. We suggest in the preceding edition of this book it can be attributed to an entity with God-like properties. We explore these properties in "God Theory" and show they predict that the Cosmos exists forever although individual universes (or incarnations of our universe) "come and go." Several other important results emerge from God Theory such a functionally triune God. The Unified SuperStandard Theory has many other important parts described in the Current Edition of *The Unified SuperStandard Theory* and expanded in subsequent volumes.

Blaha has had a major impact on a succession of elementary particle theories: his Ph.D. thesis (1970), and papers, showed that quantum field theory calculations to all orders in ladder approximations could not give scaling deep inelastic electron-nucleon scattering. He later showed the eigenvalue equation for the fine structure constant α in Johnson-Baker-Willey QED had a zero at $\alpha = 1$ not 1/137 by solving the Schwinger-Dyson equations to all orders in an approximation that agreed with exact results to 4th order in α thus ending interest in this theory. In 1979 at Prof. Ken Johnson's (MIT) suggestion he calculated the proton-neutron mass difference in the MIT bag model and found the result had the wrong sign reducing interest in the bag model. These results all appear in Physical Review papers. In the 2000's he repeatedly pointed out the shortcomings of SuperString theory and showed that The Standard Model's form could be derived from space-time geometry by an extension of Lorentz transformations to faster than light transformations. This deeper space-time basis greatly increases the possibility that it is part of THE fundamental theory. Recently, Blaha showed that the Weak interactions differed significantly from the Strong, electromagnetic and gravitation interactions in important respects while these interactions had similar features, and suggested that ElectroWeak theory, which is essentially a glued union of the Weak interactions and Electromagnetism, possibly modulo unknown Higgs particle features, be replaced by a unified theory of the other interactions combined with a stand-alone Weak interaction theory. Blaha also showed that, if Charmonium calculations are taken seriously, the Strong interaction coupling constant is only a factor of five larger than the electromagnetic coupling constant, and thus Strong interaction perturbation theory would make sense and yield physically meaningful results.

In graduate school (1965-71) he wrote substantial papers in elementary particles and group theory: The Inelastic E- P Structure Functions in a Gluon Model. Phys. Lett. B40:501-502,1972; Deep-Inelastic E-P Structure Functions In A Ladder Model With Spin 1/2 Nucleons, Phys.Rev. D3:510-523,1971; Continuum Contributions To The Pion Radius, Phys. Rev. 178:2167-2169,1969; Character Analysis of U(N) and SU(N), J. Math. Phys. 10, 2156 (1969); and The Calculation of the Irreducible Characters of the Symmetric Group in Terms of the Compound Characters, (Published as Blaha's Lemma in D. E. Knuth's book: *The Art of Computer Programming Vols. 1 – 4*).

In the early 1980's Blaha was also a pioneer in the development of UNIX for financial, scientific and Internet applications: benchmarked UNIX versions showing that block size was critical for UNIX performance, developing financial modeling software, starting database benchmarking comparison studies, developing Internet-like UNIX networking (1982) and developing a hybrid shell programming technique (1982) that was a precursor to the PERL programming language. He was also the manager of the AT&T ten-year future products development database. His work helped lead to commercial UNIX on computers such as Sun Micros, IBM AIX minis, and Apple computers.

In the 1980's he pioneered the development of PC Desktop Publishing on laser printers. and was nominated for three "Awards for Technical Excellence" in 1987 by PC Magazine for PC software products that he designed and developed.

Recently he has developed a theory of Megaverses – actual universes of which our universe is one – with quantum particle-like properties based on the Wheeler-DeWitt equation of Quantum Gravity. He has developed a theory of a baryonic force, which had been conjectured many years ago, and estimated the strength of the force based on discrepancies in measurements of the gravitational constant G. This force, operative in D-dimensional space, can be used to escape from our universe in "uniships" which are the equivalent of the faster-than-light starships proposed in the author's earlier books. Thus travel to other universes, as well as to other stars is possible.

Blaha also considered the complexified Wheeler-DeWitt equation and showed that its limitation to real-valued coordinates and metrics generated a Cosmological Constant in the Einstein equations.

The author has also recently written a series of books on the serious problems of the United States and their solution as well as a book on the decline of Mankind that will follow from current social and genetic trends in Mankind.

In the past twelve years Dr. Blaha has written over 40 books on a wide range of topics. Some recent major works are: *From Asynchronous Logic to The Standard Model to Superflight to the Stars, All the Universe!, SuperCivilizations: Civilizations as Superorganisms, America's Future: an Islamic Surge, ISIS, al Qaeda, World Epidemics, Ukraine, Russia-China Pact, US Leadership Crisis,The Rises and Falls of Man – Destiny – 3000 AD: New Support for a Superorganism MACRO-THEORY of CIVILIZATIONS From CURRENT WORLD TRENDS and NEW Peruvian, Pre-Mayan, Mayan, Anatolian, and Early Egyptian Data, with a Projection to 3000 AD,* and *Mankind in Decline: Genetic Disasters, Human-Animal Hybrids, Overpopulation, Pollution, Global Warming, Food and Water Shortages, Desertification, Poverty, Rising Violence, Genocide, Epidemics, Wars, Leadership Failure.*

He has taught approximately 4,000 students in undergraduate, graduate, and postgraduate corporate education courses primarily in major universities, and large companies and government agencies.